THE ATLANTIC AVENUE TUNNEL

THE WORLD'S OLDEST SUBWAY
1844

A Project Of
The Brooklyn Historic Railway Assn. (BHRA)
599 East 7 Street, Ste 5A
Brooklyn, NY 11218

Bob Diamond, President
Rdiamond@brooklynrail.net
718-552-7048

Copyright 1980- 2018 BHRA

Atlantic Avenue Tunnel Project Prospectus

Version 6a, October 25, 2011

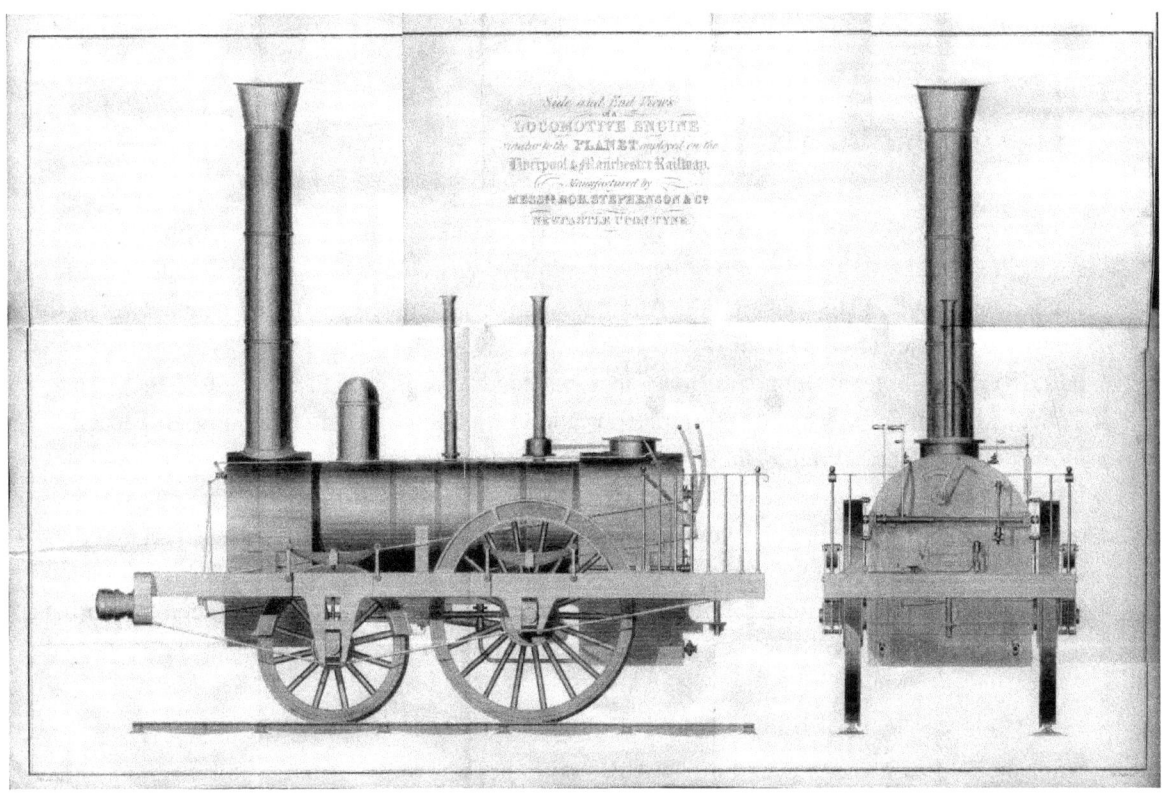

PROJECT OBJECTIVE

A very exciting proposal is now being put forth by the Brooklyn Historic Railway Association (BHRA), a non-profit education corporation. The idea is to reopen the historic Atlantic Avenue Tunnel, the oldest subway tunnel in the world, built in 1844, improving public access and restoring the tunnel as a museum and historic attraction.

BRIEF HISTORY OF THE TUNNEL

An ordinance of the Brooklyn Common Council dated March 29, 1844, granted authority to the Long Island Railroad to construct the Atlantic Avenue Tunnel. The railroad planned to use the Tunnel as a major artery in their rail service between New York and Boston. This rail line was part of a much larger system of railroads that extended from Boston to Charleston, S.C. The Tunnel was a major breakthrough in transportation technology and city planning. It carried trains under Atlantic Avenue, thereby preserving the then fashionable shopping street and its inherent pedestrian and vehicular traffic. It was the prototype of "cut and cover" subway construction, the method still used today, in which long trenches are dug in the street and then covered to form the tunnel corridors. The development of this process had an historic impact on urban planning and development; it enabled planners to integrate railroads into complex urban landscapes and led directly to the creation of metropolitan subway systems.

After the Tunnel was completed in 1844, Brooklyn became a major transportation and commercial center to rival New York, and grew to be the third largest city in the country (a distinction it held until 1898 when it became a borough of greater New York). In 1848, competition from New York in the form of the New Haven Railroad caused the LIRR to lose its monopoly on rail service to Boston, and led to substantial financial losses and the abandonment of its interstate service.

Only a few years later a prominent developer, Mr. Electus Litchfield, schemed to close the Tunnel and remove the LIRR from Brooklyn in order to create an Atlantic Avenue Boulevard and Promenade, a grandiose project inspired by the *Champs-Élysées* in Paris. With the help of corrupt politicians, Litchfield pushed the illegal legislation which permitted him to tax Atlantic Avenue merchants and property owners for the removal of the Tunnel and the LIRR, which he had branded as a "public nuisance." As a result, steam locomotives were banned in Brooklyn in 1859 and the Tunnel was finally closed and sealed in 1861. In only a few short years the Tunnel had gone from a technologically advanced project which would benefit all of Brooklyn, to a scapegoat for the corrupt plans of a robber baron. Litchfield then used the ill-gotten money to initiate his real estate project in what would become Park Slope, and build a new rail line from Jamaica to Hunters Point, the line the LIRR presently uses. However, no Boulevard was built due to the ensuing lawsuits brought by the merchants and property owners against Litchfield. The elimination of rail service left downtown Brooklyn in economic chaos, causing it to be transformed from an economic rival of New York to one of its most prized and diversified residential areas.

REDISCOVERY OF THE TUNNEL

For over one hundred years, the Atlantic Avenue Tunnel remained sealed and largely forgotten, the subject of fantastic myths and legends which seemed to crop up with each generation-- many of which turned out to have some truth. Despite the recurrent rumors, numerous attempts to locate an entrance had failed. Finally, in early 1980, Robert Diamond first heard of the legendary tunnel on a radio broadcast about *The Cosgrove Report*, which claimed there was an old steam locomotive buried in a forgotten tunnel in downtown Brooklyn. The book also mentioned a legend that the missing pages of John Wilkes Booth's diary had been hidden there. Intrigued, Diamond spent seven months researching the tunnel's history, eventually locating an unmarked manhole in the middle of Atlantic Avenue and Court Street he was sure would lead to the long-abandoned tunnel beneath. Yet when the manhole was opened, there was nothing to be seen but a three-foot drop. The dirt fill came up to about two feet from the underside of the pavement. Diamond knew at that moment he was standing on a backfilled portion of the tunnel. Looking around with a flashlight, he noticed what appeared to be a wall some seventy feet to the west. He was separated from this wall by a crawlspace less than two feet high. For the next year he searched the area, pleaded with skeptical, sometimes indifferent officials, researched, probed and slowly raised the curiosity of enough influential people to continue the exploration. In the summer of 1981, Mr. Diamond was able to crawl the seventy foot distance to the wall where he noticed the outline of a blocked-up opening in the concrete wall. The access was sealed with brick and Belgian paving blocks. After several hours of hard work with pick and shovel, Diamond and several men from Brooklyn Union Gas Company (now National Grid), who had agreed to help him on his underground mission, broke through the opening and finally saw the full expanse of the Tunnel before them, exactly as it was when sealed up 120 years earlier.

In 1982, Mr. Diamond founded a not-for-profit corporation, the Brooklyn Historic Railway Association (BHRA), to preserve and restore the tunnel, and establish a museum and scenic railway. For the past twenty-nine years, BHRA, in conjunction with elected officials, city agencies, community groups and local businesses, has been working to develop the Tunnel as a valuable public asset. BHRA received all the necessary approvals for a franchise from the City of New York to occupy and operate the tunnel as a museum devoted to the study of early railway transportation. BHRA has also been fostering public awareness and support for this forgotten municipal treasure, hosting public tours which have been enjoyed by thousands of city residents and tourists alike. During this time Mr. Diamond has further explored the tunnel's history and its significance to New York. Because it is the earliest known example of the cut-and-cover technique of railroad tunnel building in the world, and because it was part of New York's earliest train service, the tunnel has been listed on the National Register of Historic Places since 1989.

The tunnel is also recognized by the *Guinness Book of World Records* as the "World's Oldest Subway Tunnel", starting in the 2011 edition.

TUNNEL CONDITION

After being sealed for over a century, the Tunnel is a perfectly preserved, truly magnificent structure. It is a half-mile long, twenty-one feet wide and seventeen feet high. Its walls are six-foot thick granite blocks and the roof is a three-foot thick brick arch. Several prominent civil engineers have been actively engaged in determining the tunnel's structural soundness and architectural and engineering significance, and have concluded that it is structurally perfect. In fact, they have compared it to the pyramids of Egypt. An evaluation performed by LMW Engineering Group, LLC, in March 2009, found the tunnel "impressively devoid of any sign of deterioration." Their report further concludes that:

*The structural integrity of the tunnel is sound and has not been compromised by aging.

*The tunnel can be considered safe under its current use for visitors and tourist attraction.

*There is no evidence that any form of maintenance or repair work is necessary at this stage.

*The tunnel can be safely, with relatively minimum rehabilitation effort, mostly esthetic, be utilized as a museum or similar facility.

*In summary, the tunnel, as inspected by us, is a safe and sound structure.

Studies conducted by prominent consultants as well as by the City departments of Sewers, Water Supply, Transportation, Fire and Electrical Control, and a study by the National Historic Register have led to the following appraisal: The tunnel is a marvel of early engineering techniques, historically one of the most important architectural structures of the 19th century.

PUBLIC INTEREST

Since 1982, BHRA has offered both public and private tours of the tunnel which have been enjoyed by thousands of visitors. Many private and public schools have sent hundreds of students on class trips. Most recently, at the behest of the city Department of Transportation, regular public tours were reinstated in 2007 and given about twice per month through the end of 2010. During this time public interest in the tunnel and its history increased dramatically, and hundreds of people were safely led through the tunnel on guided tours given by Mr. Diamond. The Tunnel received over 12,000 visitors in 2010. Both New Yorkers and tourists from all fifty states and many foreign countries lined up for the adventure of seeing the legendary underground expanse for the first time. The response was overwhelmingly positive. Visitors reported feeling a strong sense of mystery and intrigue, as well as the sensation of travelling back in time to the 19th century. Teachers commented afterwards that students were highly motivated by the visit.

Numerous newspaper and magazine articles have been written about the Tunnel, including feature stories in The New York Times, Daily News, National Geographic, Science Digest, and The New Yorker. The project has also been covered by local television and radio as well as national exposure on CNN and the Canadian Broadcasting Corporation. In 2009, the tunnel was featured on the History Channel TV show Cities of the Underworld. In addition, National Geographic had begun work on a documentary focused on the historic locomotive buried at the western end of the tunnel.

PUBLIC BENEFITS

The Atlantic Avenue Tunnel Museum will be designed to appeal to the general public as well as to engineering, history, and architectural buffs. With its dramatic subterranean location and exhibits which will include historic train cars and railroad artifacts, the museum should prove of particular interest to children.

The museum will have both local and international appeal. The Brooklyn Historic Railway Association estimates it will draw at minimum 10,000 visitors per year during our proposed "Phase I" from the tri-state area, as well as tourists sightseeing in New York City. The museum, located beneath a busy Brooklyn thoroughfare, will also draw visitors from its immediate neighborhood, a melting pot of African Americans, Hispanics, Middle Eastern émigrés and families of Italian American descent.

Since future development in downtown Brooklyn will rely on the intrinsic assets of the area, it is the old Atlantic Avenue Tunnel which highlights the primary asset of the community—easy access and unparalleled transportation facilities. The museum, in the heart of downtown Brooklyn, is just a short walk from federal courts, office buildings, city government offices, and the historic homes of Brooklyn Heights, the first designated landmark district in the United States. The Atlantic Avenue Tunnel can thus serve as an historic symbol for today's public and private sector leaders as they reemphasize downtown Brooklyn as a business and transportation center now, as it was 160 years ago.

This project will enhance the quality of life in an area now experiencing a major renaissance, as well as ensuring the redevelopment of downtown Brooklyn from both an economic and social standpoint. It would have a synergistic impact on several other projects currently underway downtown.

As well as providing a new cultural resource and tourist attraction for the state and city, this project will stimulate business in the many restaurants, specialty food shops, antique stores, art galleries, and other retailers in the area. In addition, the project will generate a variety of jobs in its implementation, and serve as a centerpiece for the much publicized redevelopment of downtown Brooklyn.

Once accessible to the public, the Tunnel would have immediate public benefits. Current uses would include:

1. Guided walking tours to groups of up to 50 people at a time. These tours would take place on Sundays from 11:00 AM to 5:00 PM, and on Saturdays when demand warrants. Special weekday events may be planned.
2. Cultural gatherings.
3. Site location for media productions.

Possible additional future uses as per NYC Board of Estimate resolution adopted on October 9, 1986:

1. Historical exhibits.
2. Streetcar/railway museum and/or railway vehicle storage "barn."
3. Partial use as part of a future streetcar line.

PROJECT PLAN

TUNNEL DESIGN

The design work for the Atlantic Avenue Tunnel Museum will include:

Site documentation:

Gathering and obtaining scale drawings and photographs along with field measurements of existing conditions.

Schematic design:

Preparation of designs for sidewalk kiosk entrance to the tunnel at the intersection of Court Street and Atlantic Avenue; underground passage to the tunnel; underground entry hall to the museum, including location of sales office, ticket booth, and concession stand; exhibition installation within tunnel; and portal entrance and approach ramp at Columbia Street.

Presentation Drawings:

We will use existing scale plans, sections, and elevations to describe the schematic design of the project components. The final package will include scale drawings, and/or renderings, and possibly a model of the project, as needed.

PHASE I

FIRE AND SAFETY IMPROVEMENTS

Current access to the tunnel will be improved as follows:

Egress:

A second Egress will be installed in the center of Atlantic Avenue, approximately 30 feet to 60 feet west of the current manhole entrance at Court Street and Atlantic Avenue, depending on site geometry within the tunnel. This egress will come in the form of a new manhole casting and frame. There are two alternatives: a double leaf rectangular manhole casting and frame, 6 ft x 3 ft inside clearance. The second alternative utilizes a four foot inside diameter round manhole casting and frame. The installation of this new 48 inch clearance manhole, with a stair underneath, had already been approved by the DOT, in a letter to BHRA, dated November 17, 1989. See Exhibits A and B.

The extant concrete bulkhead opening near the tunnel entrance will be enlarged to a new preferred size of 78 inches x 36 inches, or as close to those dimensions as is feasible, in order to meet codes and improve access to the main body of the tunnel.

The existing wood stair will be replaced by a steel stair of similar vertical rise and angle, with a tread length of 4 feet, tread 12 inches, and 8 inch risers. A 4 ft x 4 ft steel platform will be provided at the top of the stair. Handrails of standard type will be provided along both sides of the stair and platform. The existing wooden stair will be removed, together with any other flammables. The estimated cost of this steel stair, delivered and installed, is $4,200, based upon a bid we received.

The new manhole entrance will also utilize a second steel stairway. This new second stair is partly patterned after a traditional NYC Fire Escape stair. However, OSHA now categorizes this type of stair as "Ship's Stairs." Since the current NYC Building Code is silent on new Fire Escape design, other sources were used. See Appendix, and Exhibits C and D. Subject to final design, in the first alternative a new steel stair would be utilized of approximately 11 feet (132 inches) vertical rise (providing minimum interior headroom of 80 inches), approximately 61 degree angle, tread length 36 inches, tread 6 inches, risers 9 inches, with 14 risers total. In the second alternative, a new steel stair would be utilized of approximately 11 feet vertical rise, at an angle of approximately 70 degrees, tread length 24 inches, tread 6 inches, risers 12 inches, with 11 risers total. Both alternatives would be equipped with steel handrails. However, the stair described in the second alternative will be provided with appropriate steel handrails that in cross section will be spaced a minimum of 36 inches apart. The cost of this steel stair, delivered and installed, is estimated at $3,000.

Final location of the new manhole and the second steel stair depends on obtaining exact measurements of existing tunnel geometric conditions. These measurements must be done immediately, in order to finalize this plan.

We anticipate the implementation of this plan will make the entire tunnel fire proof, and that Emergency Personnel entering the tunnel on a job will need to carry only a minimum of appropriate equipment, and traditional "gurneys" will easily fit within the tunnel.

Lighting:

We have examined § [C26-605.1] 27-381, of the NYC Building Code of 1968, as amended. We believe the existing ambient lighting within the tunnel exceeds the minimum requirement of 2 foot candles. Emergency lighting is provided by each and every visitor to the tunnel being required to furnish their own working flashlights. Visitors to the tunnel travel in specific groups, led by a long experienced guide. No "independent exploration" in the tunnel is permitted. Please note that the tunnel is only open a few hours, on a handful of days per month. It is otherwise unoccupied.

Wiring within the tunnel is of construction site type, industrial grade, consisting of insulated, weather proof and oil resistant No. 2, 3-conductor and No. 6, 3-conductor wire. All wiring connections are made to NYC Subways 3rd Rail Dept. specifications: Each connection is made with copper "bug nuts," with 3 wraps of rubber high voltage tape, 3 wraps of friction tape, and 3 wraps of PVC tape. Our wiring and generator are properly protected by appropriate circuit breakers.

An in-house electrical connection will also be provided to eliminate the need for an external generator. Hard wired Emergency lighting will be installed as required.

Communications:

A combination of cell phones and walkie-talkies will be carried by each BHRA staff person present at the tunnel. It's anticipated that once the new, enlarged Egress is installed, and the existing concrete bulkhead opening widened, radio reception inside the tunnel will be greatly improved. Landline telephone access will be installed as required.

Defibrillator:

BHRA will provide an *Elevaed* model "Life Pad Express," or equivalent. See Exhibit E. CPR certification will also be obtained for appropriate tunnel personnel.

Tunnel Event Scheduling Notice:

BHRA will provide our DOT designated contact person_____, with an advance tunnel event schedule for time periods of 60 days at a time. BHRA reserves the right to add on unanticipated additional tunnel events upon 3 working days notice to DOT.

PHASE I WORK TASK SEQUENCE:

1. Verify all measurements and dimensions by immediate site visits to the tunnel, as required.
2. Locate and mark any utilities within the planned work area. Generate and file with DOT any necessary MPT Plans for the planned work in the street.
3. Obtain Work Permits from DOT, as required.
4. Saw-cut roadway for new manhole casting, and saw-cut tunnel arch to accept manhole casting, as required. Saw-cut existing opening in concrete bulkhead to enlarge.
5. Install new manhole casting.
6. Install the new 70 degree steel stair.
7. In-load components for the new steel replacement of the existing wood stair. Dismantle existing wood stair.
8. Dispose of existing wood stair.
9. Install the replacement steel stair.

PHASE I PROJECTED COST

The projected total costs of Phase I improvements is approximately $20,000 to $25,000. Cost was based upon actual verbal bids solicited and received by BHRA, during February 2011.

PHASE II

NEW ENTRANCES AND MUSEUM

New entrances will be constructed and the tunnel will be restored as a museum and historic attraction. Project components will include:

1. Construction of one or more subway-style sidewalk entrances to the tunnel at Court Street and Atlantic Avenue, as per attached drawings. A kiosk will also be built to protect the entrance and provide shelter. One or more existing ventilation shafts will be reopened and activated to provide forced-air ventilation. Also to be installed are standpipes for fire protection and an improved museum-style lighting system.

   Estimated construction costs for these improvements is approximately $3 million. This is based upon ***Engineering News Record*** Construction Cost Indices (CCI), inflating a circa 1985 "Phase II" engineer's cost estimate, to current October 2011 dollars.

2. BHRA may also obtain the rights to access the tunnel from the basement of a suitable building on Atlantic Avenue. There is a candidate building on the SW corner of Atlantic Avenue and Clinton Street which is home to the Tripoli Restaurant. This is a very unusual structure, as it has three levels of sub-basements. There is reason to believe (subject to access to relevant City records) that the floor level of the lowest sub-basement lines up with the floor level inside the tunnel. A 19th century plan for connecting the tunnel with an entrance located in this building could be realized by soliciting the assistance of the local "Sand Hog" union as well as the expertise of one of the many coal mine construction firms of Pennsylvania and West Virginia. Of course, the appropriate written consent of the building owner would have to be obtained, which would clearly entail some form of remuneration for the use of the building. It is estimated, subject to actual bid solicitations, that this thirty-foot long pedestrian connecting tunnel could be driven just below existing utilities, right from the basement into the tunnel, within a total project budget of $750,000. Any one of many local concrete saw cutting firms, such as the J.P. Hogan firm, could readily cut through the tunnel's stone wall with relative ease, using a diamond blade hydraulic powered chain saw, or by the use of a large diameter, electric powered, wall mounted, diamond blade circular saw concrete wall cutting system that could be readily set up inside the tunnel. If this plan were implemented, the improved entrance and steel stair already proposed for the middle of Atlantic Avenue just west of Court Street would then serve as the tunnel's Emergency Exit.

3. Construction of a museum within the tunnel. Exhibits will highlight the impact the tunnel had on the economic and social development (Transit Oriented Development) of Brooklyn; in particular, the reason it was built, how it was built and why it was closed. Various eclectic scientific and historical principles, and cutting edge theories relating to rail transit science and local history will also be included in the educational presentation

within the tunnel. The centerpiece of the museum will be the circa-1830's steam locomotive discovered in the tunnel. Other exhibits will include artifacts from the tunnel and various media illustrating the tunnel in use, and Brooklyn in general during that period of time. Another major attraction will be a fully restored antique streetcar which people can ride from one end of the tunnel to the other.

Phase IIA

Phase IIA was a plan BHRA developed circa 1990, to make the early 19th century steam locomotive said to have been buried behind a tunnel wall, a major feature of the overall tunnel tour/museum experience. At that time, a method was devised to drive an approximately 60 foot long "connecting tunnel" between the buried locomotive, and the interior of the tunnel under Atlantic Avenue and Hicks Street. This work is anticipated to be accomplished without any contact with existing underground utilities, through the use of pure tunneling, and by not employing the "cut and cover method" at all.

A similar connecting tunnel, of less than half the length cited above, could be used to connect the tunnel's interior with a suitable sub-basement along Atlantic Avenue. See Appendix IIA preliminary design documents attached below.

REVENUE PROJECTION

On any typical circa 2010 Sunday afternoon tunnel tour date, regardless of season or weather conditions, BHRA received on average, about $5,000 in free will contributions ($4,000 low, and over $6,000 high). At a $15 per person suggested contribution, this translates to approximately 330 visitors per day. Visitor surveys taken over several days indicated roughly 80% of tunnel visitors ate in Atlantic Avenue restaurants, and/or patronized local shops as a direct result of the tunnel tours.

We base our future Phase II- IIA Revenue Projection upon past performance over the last three years, and the assumption that the suggested contribution for tunnel tours will be raised to $20 per person (a 30% increase), and that the planned Phase II-IIA improvements would allow the operation of tunnel tours/museum to be expanded to 7 days a week, with the circa-1830's locomotive discovered in the tunnel made part of the exhibit. Based upon the foregoing, we project "Phase II- IIA" gross revenue would be in the neighborhood of:

$6,500 per day x 360 days = $2,340,000 per year (maximum possible gross project revenue)

To be conservative, we cut this figure in half, which yields an approximate gross income of $1,170,000 per year.

PHASE III

OPENING THE WESTERN TUNNEL PORTAL- Long Term Planning

There appears to be significant community interest in the future construction of a downtown Brooklyn streetcar system, based upon certain cost effective technology for the purposes of fostering "Transit Oriented Development". The tunnel may well be determined to be an asset in the development of such a transportation improvement program, as a "trolley barn" to house the streetcars when not in use.

Circa 1989, BHRA working closely with the Brooklyn Borough President's Topographical Office, and with the NYC Department of Transportation, Department of Highways, developed a set of "Builder's Pavement Plans" (NYC DOT Plan # BNP 88-262) for the implementation of the re-opening of the tunnel's extant western portal at Columbia Street.

Approximated October 2011 cost, based upon a circa 1989 Cost Estimate, and the *Engineering News Record* "CCI" tables: $5.56 million.

APPENDICES:

Phase I

A- Double leaf manhole casting detail, 6ft x 3ft, March 2011, 2 pages

B- Circa November 17, 1989 letter from George Holuka, P.E. (Chief, NYC DOT Highway Design) and a circa September 28, 1988 letter from Dr. Michael Horodniceanu (Second Deputy Commissioner, NYC DOT) to BHRA, stating that DOT gives it permission for the current manhole to be replaced by a larger manhole, and that the DOT itself would provide a painted in pedestrian safety island around the new manhole. Three pages, and a separate plan view drawing, which had been prepared by DOT at that time.

C- **Safety and Survival on the Fire Ground,** by Vincent Dunn, 1992, Published by Fire Engineering Books, pg 261, 2 pages

D- **The Tenement House Laws of the City of New York**, Published by the City Of New York, 1903, pg 5, 2 pages, and **Ship's Stair Design Description** (including OSHA interpretation letter, dated 2/10/06), FS Industries, 2011, 5 pages

E- Preferred defibrillator unit, manufacturer's description sheet, 1 page

F- Flyers of candidate concrete wall cutting firms

G- Extracts of the circa 2008 NYC DOT tunnel consent renewal, highlighting certain key errors and other defects contained therein

H- Circa 2009 consent modification made by NYC DOT, to our circa 2008 tunnel consent renewal, requiring the creation and implementation of "MPT Plans" at the sole cost of BHRA. Letter from Emma Berenblit Director of DOT Consents, to BHRA dated July 22, 2009 and the executed Consent Modification document, dated September 9, 2009.

I- Phase 1 design documents prepared by LMW Engineering Group, June 2011. Three sheets.

J- Circa 1916 scale engineering drawings made by the City of New York

K- Approved NYC Board of Estimate Resolution, Calendar # 47, October 9, 1986

L- Tunnel safety report issued by LMW Engineering Group, March 2009

M- Building Code, City of New York, 1968, Title C Part 1, "Building Construction", § [C26-10.0]; Inapplicability of the NYC Building Code to tunnels or subways. The BHRA tunnel project, Phase I- II inclusive, was defacto "permitted" by a vote of the NYC Board of Estimate on October 9, 2011, Calendar No. 47, and by a vote of the NYC Planning Commission, prior to July 1, 2008. The tunnel project is therefore "grandfathered in" under the aegis of the original NYC Building Code of 1968. Reference source: NYC Building Code, as revised July 1, 2008, Preface Section, page IIB. Needless to say, the current (July 1, 2008) NYC Building Code will be strictly adhered to where ever technically feasible, given the unique nature of the tunnel site.

Phase II
1). Complete "Plans, Specification & Estimates" package (PS&E) prepared by Steven Carroll, P.E. circa 1985.

2). *Engineering News Record* "CCI" tables, 1978- Oct 2011

Phase IIA
Connecting tunnel design concepts, circa 1990

Phase III (Re-Opening of Original Western Tunnel Portal Near Columbia Street)

1). Completed circa 1989 NYC DOT Builder's Pavement Plan # BNP 88-262. Three sheets.

2.) Circa June 9, 1988 letter from Bob Diamond (BHRA) to NYC DOT Commissioner Ross Sandler

3.) Circa July 25, 1988 letter from Thomas Markham, PE. (NYC DOT) to Bob Diamond (BHRA)

4.) Circa January 13, 1989 meeting letter from Gerard Renninger, P.E. (NYCDOT)

5.) Circa February 9, 1989 meeting minutes (NYCDOT, State DOT, BHRA)

6.) Circa March 9, 1989 letter from Anthony Cosentino, P.E. (NYCDOT)

This Space Left Intentionally Blank

Long Island Rail-Road Office,
27th Nov. 1844.

The Directors of the Long Island Rail-Road Company request the favour of your attendance at Brooklyn, on *Tuesday* the 3 of December, on the occasion of opening their Tunnel for use.

The Trains will leave the Depot at the South Ferry at 12 o'clock, noon.

Should the weather prove unfavourable, the ceremony will be postponed to the first fair day.

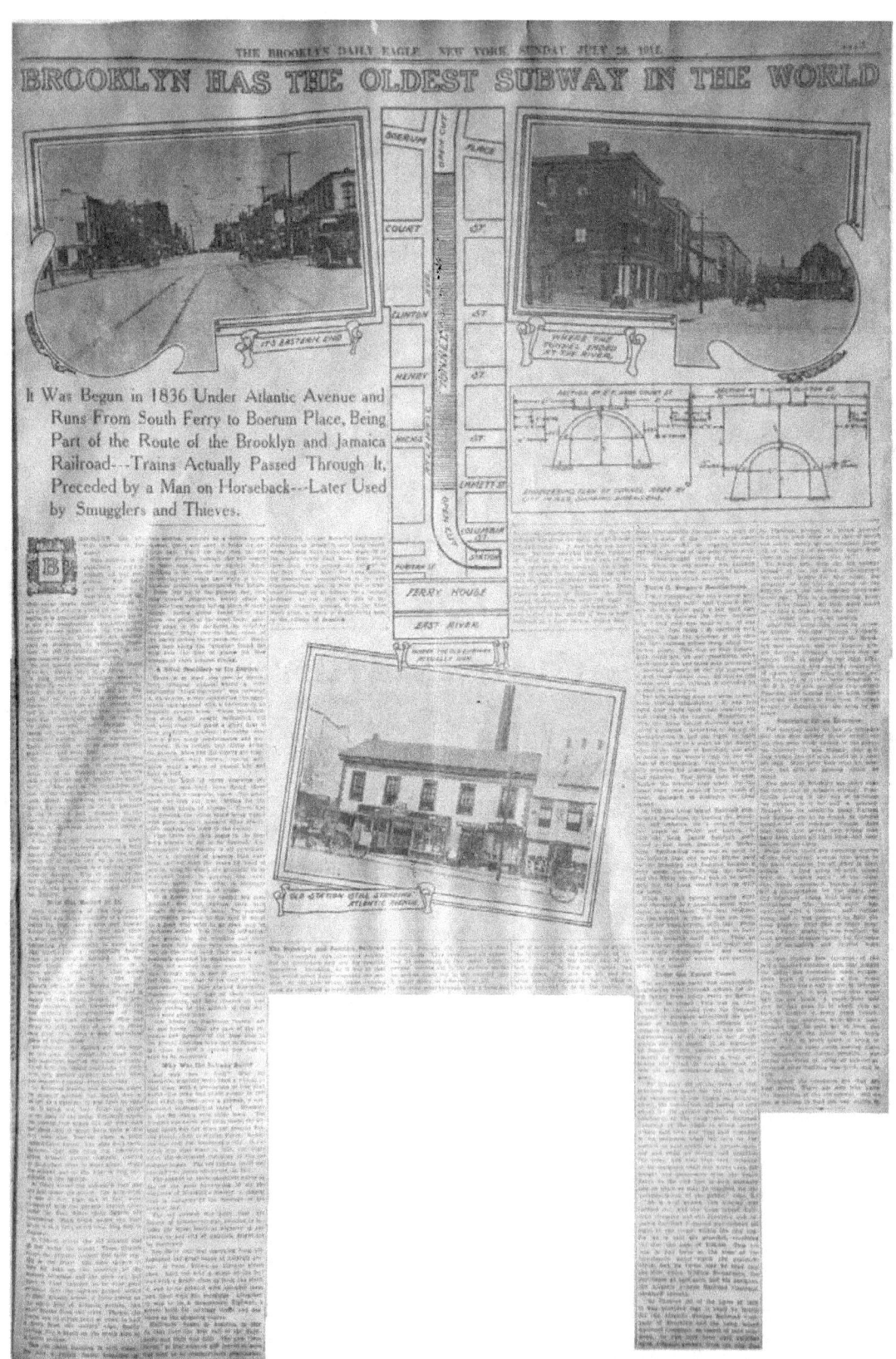

June 28, 1916

THE BROOKLYN DAILY EAGLE, NEW YORK, SUNDAY, JUNE 28, 19__

RAZING LANDMARK ON ATLANTIC AVE.

Old Terminal of Brooklyn and Jamaica Railroad Coming Down.

ONCE USED AS A MUSEUM.

Built in 1836, Flimsy Old Structure Now Yields Memories and Relics.

OLD ATLANTIC AVENUE R. R. TERMINAL BEING DEMOLISHED

Social Events of Note

TRAFFIC IN TITLES SETS GERMANY AGOG

Revelations by Socialist Leader Show Grafting in High Circles.

HOW TO WIN PROFESSORSHIP

First Scientific Article and Pay $10,000 "in a Certain Quarter" for Honor.

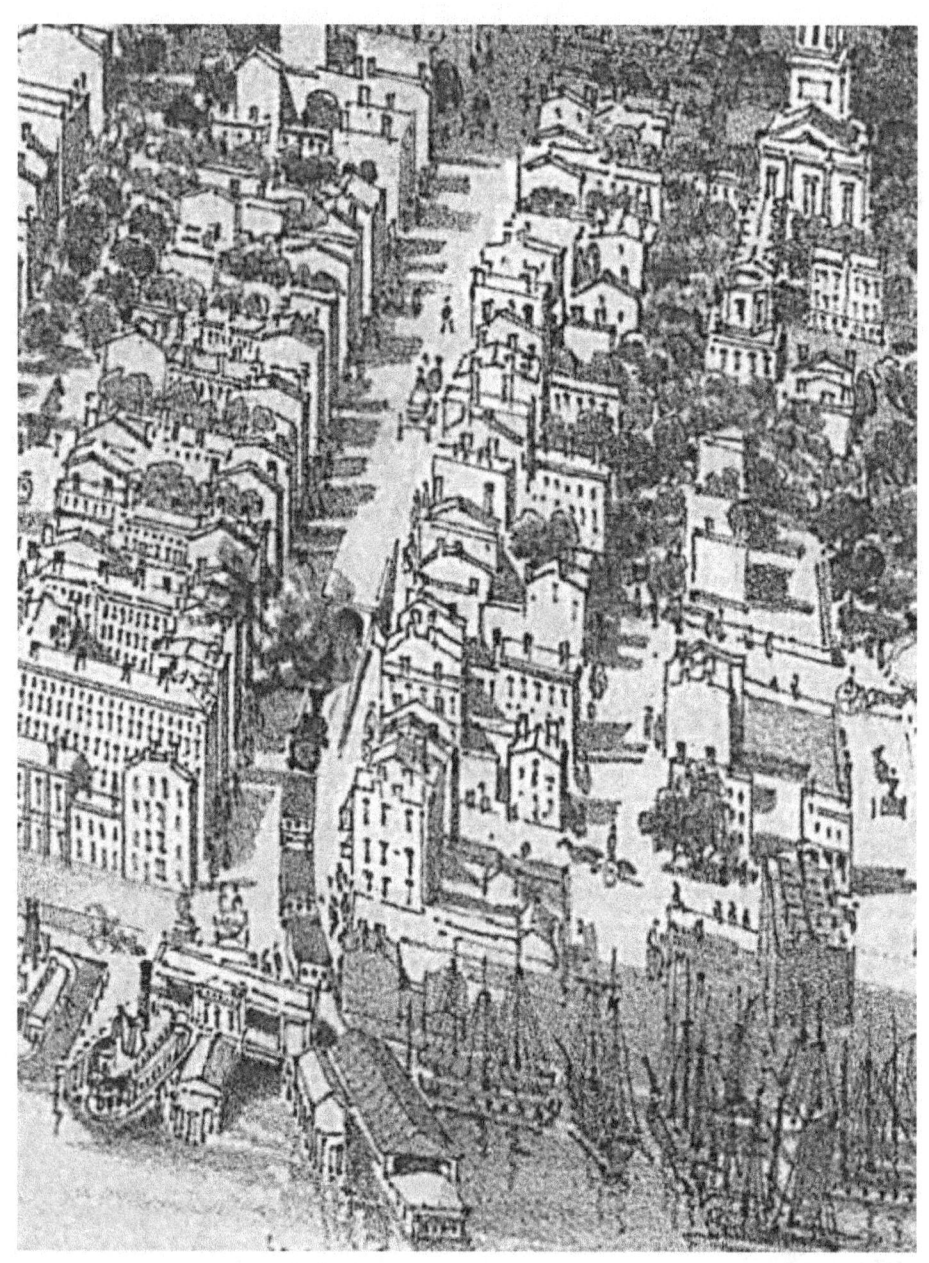

Exhibit "A"

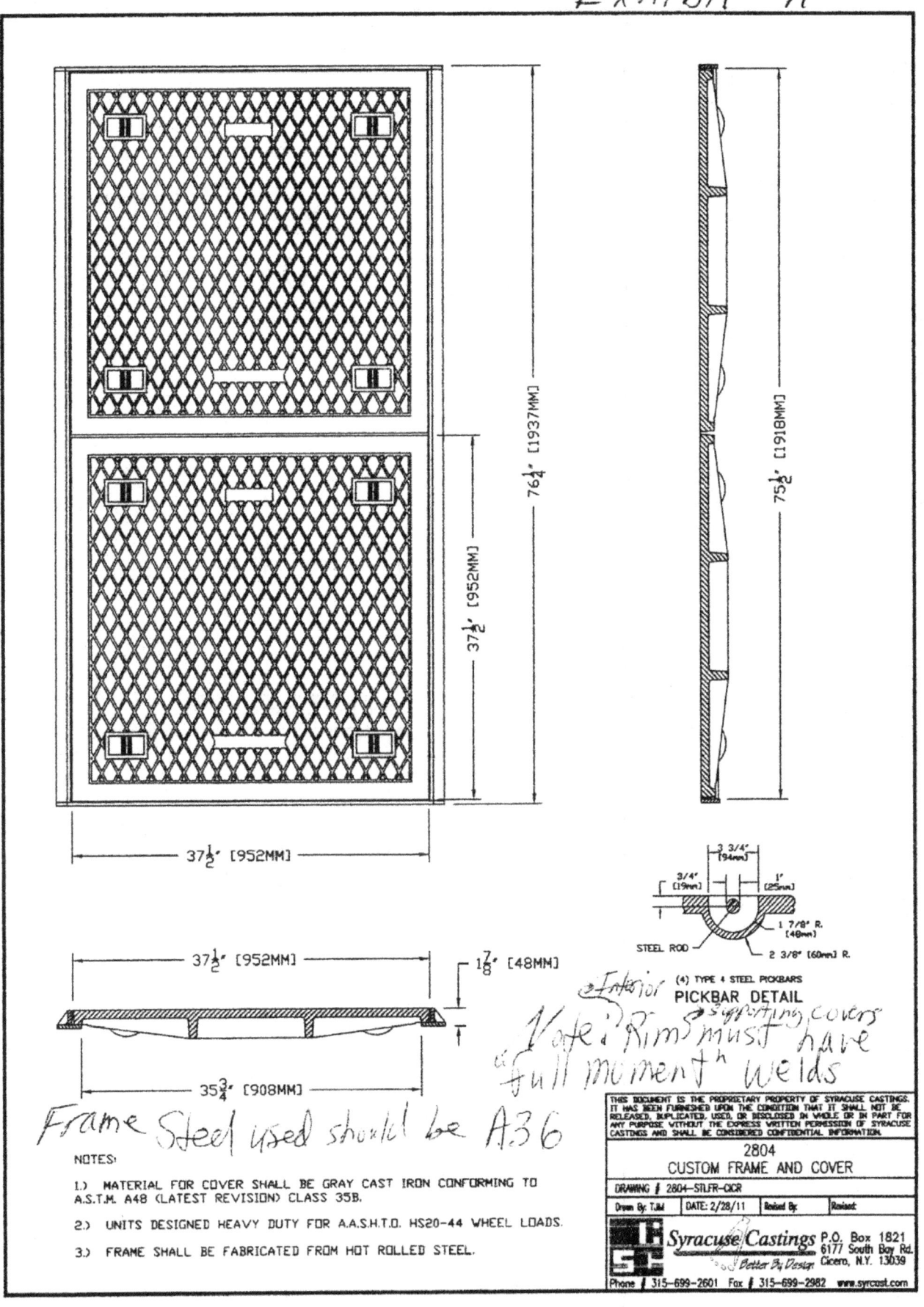

Frame Steel used should be A36

Note: Rims supporting covers must have "full moment" welds

Interior

NOTES:
1.) MATERIAL FOR COVER SHALL BE GRAY CAST IRON CONFORMING TO A.S.T.M. A48 (LATEST REVISION) CLASS 35B.
2.) UNITS DESIGNED HEAVY DUTY FOR A.A.S.H.T.O. HS20-44 WHEEL LOADS.
3.) FRAME SHALL BE FABRICATED FROM HOT ROLLED STEEL.

2804
CUSTOM FRAME AND COVER
DRAWING # 2804-STLFR-CICR
Drawn By: TJM   DATE: 2/28/11

Syracuse Castings
P.O. Box 1821
6177 South Bay Rd.
Cicero, N.Y. 13039
Phone # 315-699-2601 Fax # 315-699-2982 www.syrcast.com

From: Penny Urquhart (penny@syrcast.com)
To: rdiamond@brooklynrail.net;
Date: Tue, March 1, 2011 4:00:06 PM
Cc:
Subject: FW: Special Manhole Castings

Bob,

I have attached a drawing of what I hope is what you are requiring.

Price per unit = Frame & (2) Covers @ $1,443.00/SET.

Delivery to be determined when address is supplied.

Thanks, Penny

From: Robert Diamond [mailto:rdiamond@brooklynrail.net]
Sent: Friday, February 25, 2011 4:52 PM
To: sales@syrcast.com; sales@syrcast.com
Subject: Special Manhole Castings

Hi,
We need a doubled sized version of your 2804A, measuring a total of 72 x 36 inches, with two cover leaf castings of 36 x 36 each. Can you supply, and how much would it cost?
Thanks
Bob Diamond

This Email has been scanned for all viruses by PAETEC's Hosted E-mail Security Services, utilizing MessageLabs proprietary SkyScan infrastructure. For more information on a proactive anti-virus service working around the clock, around the globe, visit http://www.paetec.com.

This Email has been scanned for all viruses by PAETEC's Hosted E-mail Security Services, utilizing MessageLabs proprietary SkyScan infrastructure. For more information on a proactive anti-virus service working around the clock, around the globe, visit http://www.paetec.com.

NEW YORK CITY
DEPARTMENT OF TRANSPORTATION
Ross Sandler *Commissioner*
Samuel I. Schwartz, P.E. *First Deputy Commissioner*

BUREAU OF TRAFFIC
DIVISION OF TRAFFIC ENGINEERING
28-11 QUEENS PLAZA NORTH • L.I.C., N.Y. 11101

Exhibit "B"

Dr. Michael F. Horodniceanu
Second Deputy Commissioner

November 17, 1989

Elizabeth H. Theofan, P.E.
Assistant Commissioner

Mr. Robert Diamond
President
Brooklyn Historic Railway Assn.
599 E. 7th Street
Brooklyn, NY 11218

Re: BN-AACRD-BLG-HWK197F
BT 89-2225
Atlantic Ave. Railway
Tunnel, Bklyn.

Dear Mr. Diamond:

We have reviewed your request concerning enlarging the existing access into the tunnel at Atlantic Avenue and Court Street, and have the following comments:

The center of the staircase opening should be on the centerline of Atlantic Avenue, or as close as possible.

The length (4 ft. dimension) should start at the East edge of the existing manhole, and go West toward Clinton Street.

If these two comments are included into your final plan for the tunnel access, our approval for this location is hereby given.

The roadway casting and staircase should be in compliance with Bureau of Highways specification for roadway grating.

Upon receiving the Bureau of Highways approval, you should submit your plan to the MTCCC for the specifications and permit for working in the intersection of Atlantic Avenue and Court Street.

2.

Feel free to contact me at (718) 830-7511 if you have a question regarding this matter.

Very truly yours,

GEORGE HOLUKA
Chief, Division of Highway Design

GH:js/sw

cc: D/C David Gurin
    A/C Barney La Greca, P.E.
    D/A/C P. Kaneshiro
    M. Benson
    K. Keegan
    File

NEW YORK CITY
DEPARTMENT OF TRANSPORTATION
BUREAU OF TRAFFIC

28-11 QUEENS PLAZA NORTH L.I.C., N.Y. 11101

Ross Sandler
Commissioner

September 28, 1988

Dr. Michael F. Horodniceanu
Deputy Commissioner

Samuel I. Schwartz, P.E.
Chief Engineer First Deputy Commissioner

Mr. Robert Diamond
President
Brooklyn Historic Railway Assn.
509 E. 7 Street
Brooklyn, NY 11218

Re: DOT 053881
BT 89156
881517

Dear Mr. Diamond:

This is in response to your letter to Commissioner Ross Sandler concerning the historic railway tunnel entrance at Atlantic Avenue and Court Street, Brooklyn.

Your request has been reviewed by the Highway Design Division of the Traffic Bureau. It has been determined that a channelization could be designed which would safely divert vehicles around the tunnel entrance while not adversely affecting traffic flow.

As Mr. Holuka of the Highway Design division previously informed you, the installation of this painted island will necessitate the removal of a number of parking meters. Since your discussion with Mr. Holuka indicates that the historic railway tunnel will receive limited usage during the winter months, the work required for installation has been scheduled for the spring of 1989.

Please contact Mr. Holuka at least one month before the active use of the tunnel entrance begins, so that the various phases of the work may be properly coordinated.

Very truly yours,

Dr. Michael F. Horodniceanu
Second Deputy Commissioner

MFH:mpd/sw

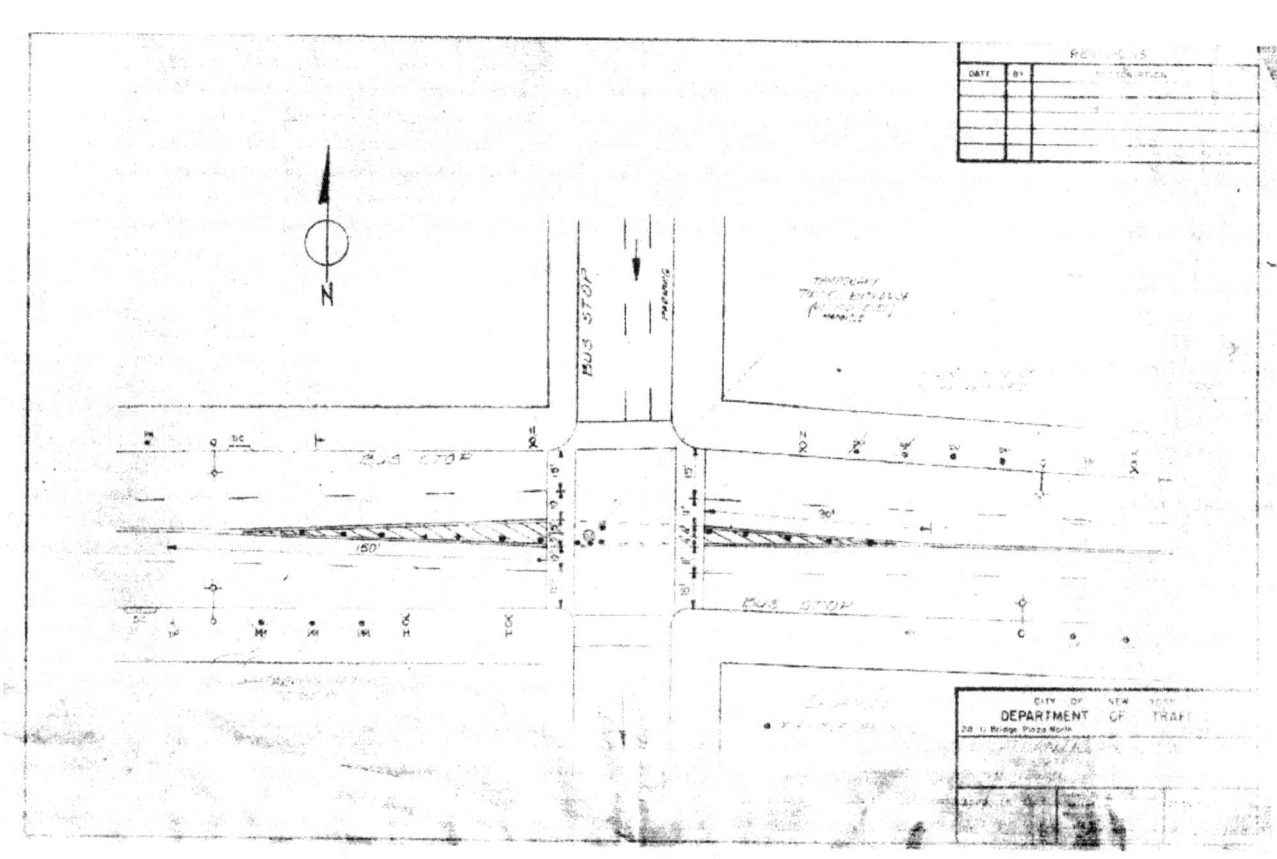

Exhibit "C"

Angle of Fire Escape Ladder

A hidden danger in a standard fire escape is the angle of its stairway and ladders. The climbing angles of a gooseneck ladder, drop ladder, and intermediate stair between fire escape balconies are much steeper than that of a normal building stairway. To compensate for these larger angles, exercise greater caution when ascending or descending a fire escape.

A typical stairway in a building rises at a 30- or 45-degree angle from the horizontal floor level. A standard fire escape stairway rises at a 60- or 75-degree angle from the horizontal floor level; the gooseneck and drop ladder rise straight up, at a 90-degree angle.

The SCBA on a firefighter's back changes his center of gravity—there is actually a constant force pulling the firefighter backward (Figure 15.17). He must be conscious of this force at all times during the climb up or down a 90-degree-angle gooseneck or drop ladder. Momentary release of a grip when climbing hand over hand on the rungs of a fire escape drop ladder will cause the firefighter to fall backward.

Figure 15.17. The weight of the SCBA air tank on one's back changes a firefighter's center of gravity. The weight of the air tank can cause a firefighter to fall backward when climbing a steep stair or a ladder on a fire escape.

About the Author

VINCENT DUNN is a deputy chief serving with the New York City Fire Department—a 35-year veteran who rose through the ranks of the department: seven years a firefighter, nine years a company officer, 19 years a chief.

Attending college at night with the assistance of the G.I. Bill, he received an A.A.S., B.A., and M.A. from Queens College, City University of New York.

An adjunct professor of Manhattan College, he taught fire engineering in the civil engineering department; currently an adjunct instructor with the National Fire Academy, he developed and teaches a residence course, "Command and Control of Fire Department Major Operations."

A contributing editor with *Fire Engineering* magazine and *Firehouse* magazine, he has authored many articles on firefighting safety and survival.

He is the author of the book and video series, "Collapse of Burning Buildings."

Exhibit "D"

twelve inches apart and not less than fifteen inches in length.

(1) The fire escapes shall open directly from at least one room in each apartment at each story above the ground floor, and no fire escape shall be placed in a court except as provided in section fifty-seven of this act. Fire escapes may project into the public highway to a distance not greater than four feet beyond the building line.

(2) The fire escapes shall consist of outside open iron balconies and stairways. The stairways shall be placed at an angle of not more than sixty degrees, with steps not less than six inches in width and twenty inches in length, and with a rise of not more than nine inches. The balcony on the top floor, except in case of a front fire escape, shall be provided with a goose-neck ladder leading from said balcony to and above the roof.

(3) Balconies.—The balconies shall not be less than three feet in width, taking in at least one window of each apartment at each story above the ground floor. They shall be below and not more than one foot below the window sills and extend in front of and not less than nine inches beyond each window. There shall be a landing not less than twenty-four inches square at the head and foot of each stairway. The stairway opening on each platform shall be of a size sufficient to provide clear headway.

(4) Floors of balconies.—The floors of balconies shall be of wrought iron or steel slats not less than one and a half inches by three-eighths of an inch,

THE TENEMENT HOUSE LAWS
OF
THE CITY OF NEW YORK.

THE TENEMENT HOUSE ACT.
(Chapters 334 and 555, Laws of 1901 ; Chapter 352, Laws of 1902.)

THE GREATER NEW YORK CHARTER.
(Chapter 466, Laws of 1901.)

Prepared for the Tenement House Department.

1903.

ENGINEERED STEEL PRODUCTS
Priced Full Line Products Catalog
PO Box 72659 Providence, RI 02907
Toll free (800) 421-0314 • In RI (401) 421-0314 • Fax (401) 421-5679

Home   Index   Order   Search   Quote   Info   About Us   Services

Categories >>> Select a Product Category   Thursday, March 3, 2011

back to Product Category

SHIPS LADDER - 68° STEEP INCLINE

STRUCTURAL STEEL STAIRWAYS SHIPS LADDER DESIGN

safe access with minimum space requirement

- Quality engineered to your specific height requirement.
- Designed for safety and convenience.
- Heavy duty bar grating stair treads won't sag or dish (supplied as standard).
- Factory welded handrails of 1 1/2" x 14 ga. square tubing.
- 10" structural channel stringers.
- Fire proof construction.
- Standard finish gray enamel, others available upon request.
- Standard 24" wide treads, 27" overall stair width. Other widths available upon request.
- Tread depth is 6".
- Our standard finish is a quick dry gray enamel. Other colors available.

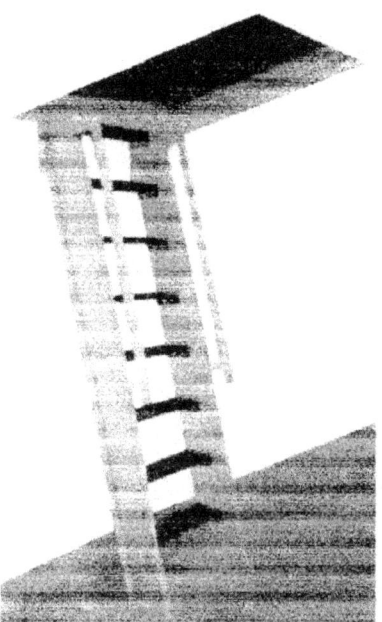

(C) Standard Extended Handrail Models

(D) Hatch Access Models (Short Handrail)

OSHA CONFORMANCE
See Details Below Price Tables

NEW! Optional Finishes!

http://www.fsindustries.com/more_info/ships_ladder/ships_ladder.shtml   3/3/2011

Now we are pleased to offer our products with a hot-dipped galvanized finish or in a duplex system which is paint over a hot-dipped galvanized undercoat.

CUSTOM SIZES - SAME DAY QUOTES

Custom sizes and configurations are readily available. We provide same day price quotes if you require a special stair angle, specific tread rise/run, or a maximum horizontal stair run to satisfy available floor space constraints.

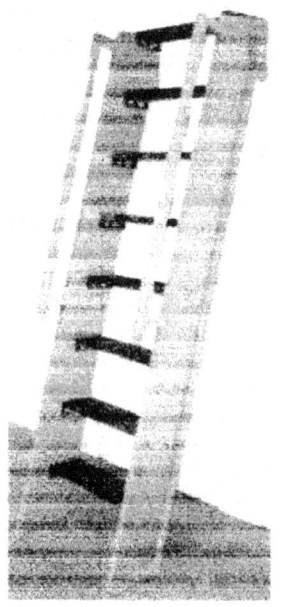

Optional Vertical Handrail to 42" guard height quoted upon request.

Optional walk-thru handrail quoted upon request.

Add $100.00 to standard pricing and add - VHR to Model No.

Add $150.00 to standard pricing and - WTHR to Model No.

Steel Stair Treads

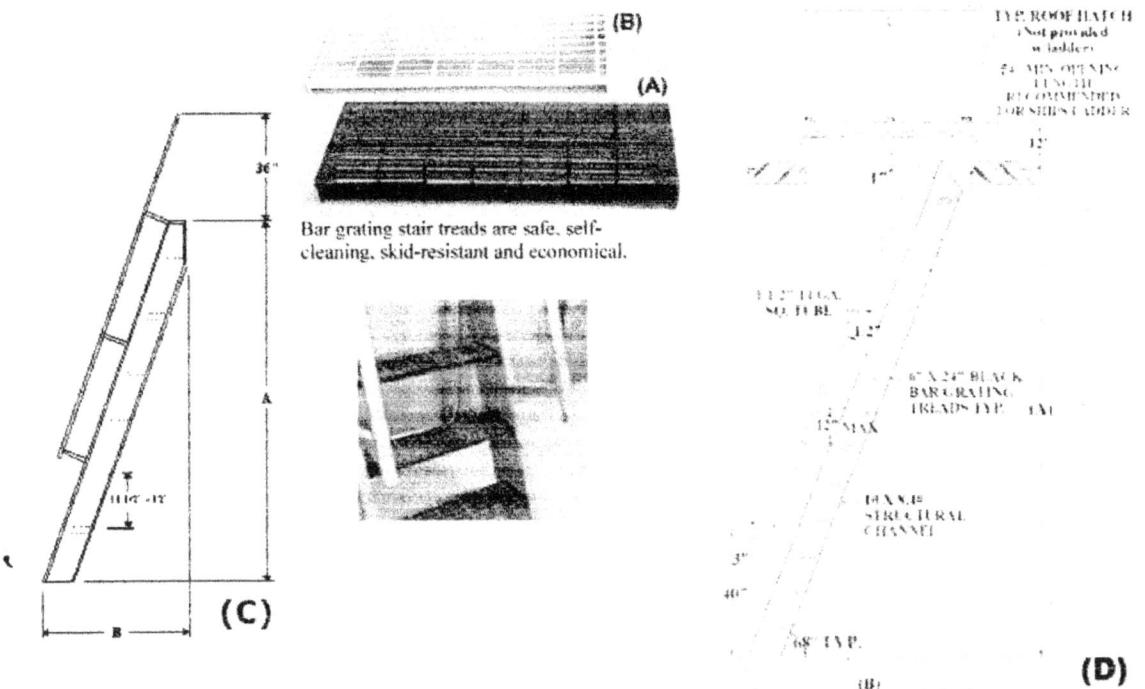

Bar grating stair treads are safe, self-cleaning, skid-resistant and economical.

Important Note When Ordering: Specify actual floor to floor height when ordering. stairways are custom designed to meet your height requirements. (A Dimension) *Other sizes available upon request.

(C) PRICING FOR STANDARD EXTENDED HANDRAIL MODELS

Type Model No.	(A) Height Range		(B) Horizontal Run	Tread Width	Nominal Overall Width	Price w/Gray Enamel & Black Bar Grating Treads (A)	Price with all Galvanized Finish (B)	Price with all Duplex Paint System
	Low Ht. (in.)	Upper Ht. (in.)						
SL2438	36"	41"	19.47" - 21.49"	24"	27"	$715	$1,199	$1,320
SL2444	42"	47"	21.90" - 23.92"	24"	27"	$798	$1,300	$1,436
SL2450	48"	53"	24.32" - 26.34"	24"	27"	$879	$1,398	$1,554
SL2456	54"	59"	26.74" - 28.76"	24"	27"	$960	$1,496	$1,669
SL2462	60"	65"	29.17" - 31.19"	24"	27"	$1,042	$1,595	$1,786
SL2468	66"	71"	31.59" - 33.61"	24"	27"	$1,124	$1,693	$1,900
SL2474	72"	77"	34.02" - 36.04"	24"	27"	$1,204	$1,792	$2,016
SL2480	78"	83"	36.44" - 38.46"	24"	27"	$1,285	$1,892	$2,132
SL2486	84"	89"	38.87" - 40.89"	24"	27"	$1,366	$1,990	$2,248
SL2490	90"	95"	41.29" - 43.31"	24"	27"	$1,447	$2,090	$2,364
SL2498	96"	101"	43.71" - 45.73"	24"	27"	$1,531	$2,187	$2,481
SL24104	102"	107"	46.14" - 48.16"	24"	27"	$1,612	$2,287	$2,597
SL24110	108"	113"	48.56" - 50.58"	24"	27"	$1,693	$2,385	$2,714
SL24116	114"	119"	50.99" - 53.01"	24"	27"	$1,773	$2,484	$2,828
SL24122	120"	125"	53.41" - 55.43"	24"	27"	$1,857	$2,582	$2,944
SL24128	126"	131"	55.83" - 57.85"	24"	27"	$1,939	$2,681	$3,060
SL24134	132"	137"	58.26" - 60.28"	24"	27"	$2,021	$2,780	$3,177
SL24140	138"	143"	60.68" - 62.70"	24"	27"	$2,102	$2,877	$3,294
SL24146	144"	149"	63.11" - 65.13"	24"	27"	$2,184	$2,977	$3,410
SL24152	150"	155"	65.53" - 67.55"	24"	27"	$2,264	$3,075	$3,526
SL24158	156"	161"	67.95" - 69.97"	24"	27"	$2,346	$3,174	$3,642

Model No.	Lower Ht.	Upper Ht.	Horizontal Run	Tread Width	Nominal Overall Width	Price A	Price B	Price C
SL24164	162"	167"	70.38" - 72.40"	24"	27"	$2,425	$3,273	$3,758
SL24170	168"	174"	72.80" - 75.23"	24"	27"	$2,508	$3,371	$3,874
SL24178	174"	179"	75.23" - 77.25"	24"	27"	$2,590	$3,470	$3,991
SL24184	180"	185"	77.65" - 79.67"	24"	27"	$2,670	$3,568	$4,107
SL24190	186"	191"	80.08" - 82.10"	24"	27"	$2,752	$3,667	$4,222

(D) PRICES FOR FLUSH TREAD HATCH ACCESS MODELS

Type Model No.	(A) Height Range Lower Ht. (in)	(A) Height Range Upper Ht. (in)	(B) Horizontal Run	Tread Width	Nominal Overall Width	Price with Gray Enamel & Black Bar Grating Treads (A)	Price with all Galvanized Finish (B)	Price with all Duplex Paint System
FTHASL2438	36"	41"	24.32" - 26.34"	24"	27"	$1,022	$1,748	$1,929
FTHASL2444	42"	47"	26.74" - 28.76"	24"	27"	$1,058	$1,752	$1,941
FTHASL2450	48"	53"	29.17" - 31.19"	24"	27"	$1,125	$1,819	$2,025
FTHASL2456	54"	59"	31.59" - 33.61"	24"	27"	$1,183	$1,868	$2,088
FTHASL2462	60"	65"	34.02" - 36.04"	24"	27"	$1,259	$1,951	$2,189
FTHASL2468	66"	71"	36.44" - 38.46"	24"	27"	$1,326	$2,022	$2,270
FTHASL2474	72"	77"	38.86" - 40.88"	24"	27"	$1,404	$2,108	$2,377
FTHASL2480	78"	83"	41.29" - 43.31"	24"	27"	$1,473	$2,189	$2,473
FTHASL2486	84"	89"	43.71" - 45.73"	24"	27"	$1,554	$2,281	$2,583
FTHASL2490	90"	95"	46.14" - 48.16"	24"	27"	$1,626	$2,366	$2,683
FTHASL2498	96"	101"	48.56" - 50.58"	24"	27"	$1,710	$2,461	$2,795
FTHASL24104	102"	107"	50.99" - 53.01"	24"	27"	$1,786	$2,549	$2,899
FTHASL24110	108"	113"	53.41" - 55.43"	24"	27"	$1,865	$2,645	$3,015
FTHASL24116	114"	119"	55.83" - 57.85"	24"	27"	$1,943	$2,735	$3,118
FTHASL24122	120"	125"	58.26" - 60.28"	24"	27"	$2,026	$2,830	$3,233
FTHASL24128	126"	131"	60.68" - 62.70"	24"	27"	$2,102	$2,922	$3,341
FTHASL24134	132"	137"	63.11" - 65.13"	24"	27"	$2,185	$3,020	$3,457
FTHASL24140	138"	143"	65.53" - 67.55"	24"	27"	$2,263	$3,111	$3,566
FTHASL24146	144"	149"	67.95" - 69.97"	24"	27"	$2,344	$3,211	$3,682
FTHASL24152	150"	155"	70.38" - 72.40"	24"	27"	$2,422	$3,303	$3,792
FTHASL24158	156"	161"	72.80" - 74.82"	24"	27"	$2,505	$3,356	$3,909
FTHASL24164	162"	167"	75.23" - 77.25"	24"	27"	$2,581	$3,495	$4,020
FTHASL24170	168"	174"	77.65" - 80.08"	24"	27"	$2,664	$3,593	$4,135
FTHASL24178	174"	179"	80.08" - 82.10"	24"	27"	$2,743	$3,688	$4,247
FTHASL24184	180"	185"	82.50" - 84.52"	24"	27"	$2,824	$3,786	$4,363
FTHASL24190	186"	191"	84.92" - 86.94"	24"	27"	$2,903	$3,882	$4,474

SAFETY WARNING: These ladders are to be ascended and descended with the user always facing the ladder. Never descend the ladder facing away from the ladder.

OSHA CONFORMANCE

A frequently asked question is whether a ships ladder conforms to OSHA. The answer is somewhat complicated. There is no single OSHA standard which specifically relates to the ships ladder design shown here. This product is a hybrid which is neither a stair nor ladder and therefore has dimensions and design parameters which overlap and/or conflict with OSHA standards for fixed stairways (Standard 1917.120) and fixed ladders (Standard 1910.27). Also included below is a reprint of a standard interpretation letter from OSHA dated 2/10/2006.

Does this product meet OSHA requirements? We believe the answer is yes when: restricted areas preclude alternatives and a due diligence safety review of the intended use and application has been performed by the end user. These ships ladders are not intended to replace applications which require regular stairways but rather to fulfill the needs created by restricted areas. Responsibility for determining the suitability of a particular use and application rests with the purchaser.

STANDARD INTERPRETATIONS

02/10/2006 - The use of ship's stairs instead of fixed stairs in general industry.

• Standard Number: 1910.27

February 10, 2006

Ms. Erin Flory
Indiana Department of Labor
402 W. Washington Street, Room W195
Indianapolis, IN 46204

Dear Ms. Flory:

Thank you for your July 20, 2005, fax to the Occupational Safety and Health Administration (OSHA). Your letter was forwarded to the Directorate of Enforcement Program's (DEP's) Office of General Industry Enforcement for response. This letter constitutes OSHA's interpretation only of the requirements discussed and may not be applicable to any questions not delineated within your original correspondence. You had specific questions regarding the use of ship's stairs in general industry.

The questions below have been restated for clarity.

Question 1: Are ship's ladders (also known as ship's stairs) required to meet the fixed ladder requirements in 29 CFR 1910.27?

Response: No. The standards for fixed ladders in §1910.27 do not apply to ship's stairs.

Question 2: Can ship's stairs be used in general industry?

Response: Existing §1910.27 does not address ship's stairs. However, the 1990 Proposed Rule for Subpart D, Walking and Working Surfaces, 55 Federal Register 13360, addresses this issue at proposed §1910.25, Stairs, paragraph (e)(1), which states, "Ship's stairs shall be installed at a slope between 50 degrees and 70 degrees from the horizontal." Where an employer is in compliance with the provisions of a proposed standard, it is OSHA's general policy to treat the violation of an existing requirement as a de minimis violation. Therefore, in areas where conventional industrial stairs cannot be installed due to limited space availability, then OSHA would consider the installation of fixed industrial stairs with a slope between 50 degrees to 70 degrees from the horizontal to be a de minimis violation. De minimis violations are violations of standards which have no direct or immediate relationship to safety or health, and do not result in a citation, or penalty and need not be abated.

Thank you for your interest in occupational safety and health. We hope you find this information helpful. OSHA requirements are set by statute, standards, and regulations. Our interpretation letters explain these requirements and how they apply to particular circumstances, but they cannot create additional employer obligations. This letter constitutes OSHA's interpretation of the requirements discussed. Note that our enforcement guidance may be affected by changes to OSHA rules. Also, from time to time we update our guidance in response to new information. To keep apprised of such developments, you can consult OSHA's website at http://www.osha.gov. If you have any further questions, please feel free to contact the Office of General Industry Enforcement at (202) 693-1850.

Sincerely,

Richard E. Fairfax, Director
Directorate of Enforcement Programs

Hatch Access, Roof Access Ladders, Ladders, Ships Ladder, 68° Steep Incline, Structural Steel Stairways Ships Ladder Design, Access with Minimum Space Requirement, Bar Grating Stair Treads, Welded Handrails, Fire Proof Construction, and Galvanized Stairs from your complete source for material handling equipment.

Back to Product Category

http://www.fsindustries.com/more_info/ships_ladder/ships_ladder.shtml   3/3/2011

Exhibit "E"

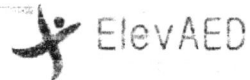

Shopping Cart
0 item(s) in cart/ total: $0.00

| Home | Website | About Us | My Account | Contact Us |

search by keyword

Products
AED Packages
AED Accessories
Cabinets & Enclosures
Batteries & Electrodes
Carrying Cases
Signs, Decals

Home > AED Packages
LifePAD Express
Part Number LifePAD_Express

Price
Your Price: $1,345.00

 SHIPPING

Share |

Choose Options

Choose
Carrying   With Carrying Case [+$50.00]
Case

Quantity 1   Add to Cart

Description
LIFEPAD© EXPRESS Package includes:

- **PHYSIO-CONTROL LIFEPAK® EXPRESS** - with one set of adult electrode pads, battery, and quick reference card. Brochure
- **Compact Stainless Wall Cabinet** - *All weather* surface mount corrosion resistant 304 stainless steel enclosure.
- Windowed, *lockable*, polyurethane *gasketed* door resists moisture, dust. Rain gutter.
- Wireless alarm has *3 levels* - 110 dB, chime, or off.
- *Genuine* Physio-Control AED & CPR responder kit.
- Two "AED on-premises" window decals
- *Dial-out capability* with alarm host kit. ($95 option)

Five year warranty on all components.

Related Items

Defibrillator

http://elevaed.3dcartstores.com/LifePAD-Express_p_57.html   2/26/2011

Exhibit "F"

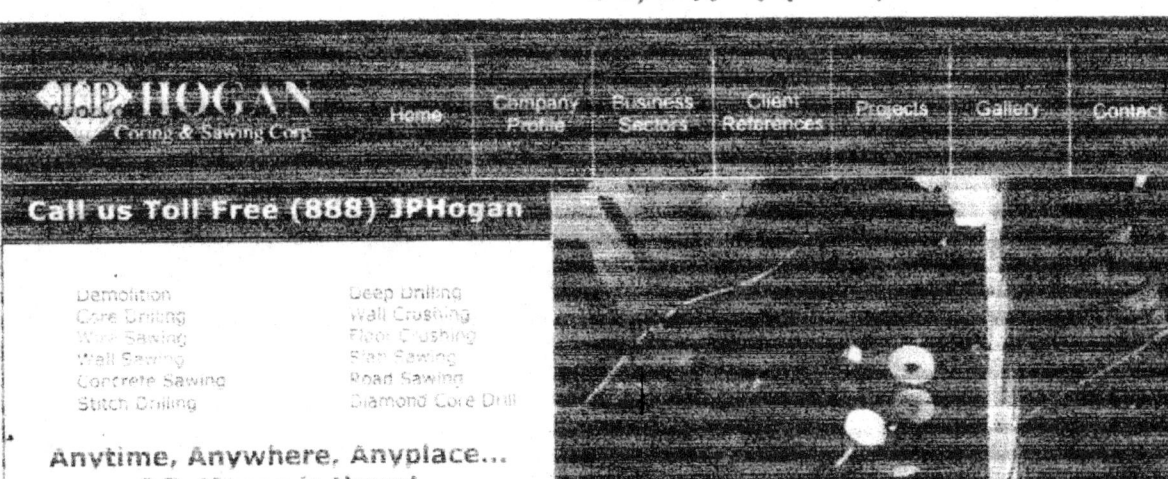

Call us Toll Free (888) JPHogan

- Demolition
- Core Drilling
- Wire Sawing
- Wall Sawing
- Concrete Sawing
- Stitch Drilling
- Deep Drilling
- Wall Crushing
- Floor Crushing
- Slab Sawing
- Road Sawing
- Diamond Core Drill

**Anytime, Anywhere, Anyplace...
J.P. Hogan is there!**

Latest News

J.P. Hogan now offers a wide array of Demolition Services.

Our expert method enables you to retain the old facade while renovating the inside of buildings and demolishing concrete with a concrete crusher, day or night.

We also provide recycling of the material resulting from the demolition, which must be disassembled into its component parts and sorted. For a complete overview of our Demolition Services, please click here.

Project Showcase

Battery Park Project
Location: Battery Park, Southern Tip of Manhattan

Problem: Underwater wire sawing for access to tunnels beneath Battery Park. This Procedure required  setup during tidal shifts. It also required the use of experienced divers.

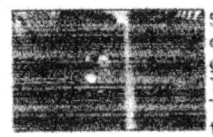 Skilled technicians drill drill into a granite foundation. To the right, Technicians mount the wire pulleys for guiding and routing the diamond wire.

Featured Services

 Demolition
Expert Demolition concrete crushing & removal services.

 **Core Drilling**
Cut smooth clean holes quickly and easily, up to a 96" diameter.

 **Wall Sawing**
Silent, Dustless, Vibration-Free Cutting. Zero structural damage.

 **Slab Sawing**
Clean, precise, & dust free. Depths up to 27 inches.

 **Wire Sawing**
Remove large dimensions of concrete easily & effectively.

Affiliations

 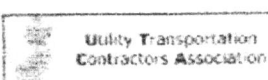  

Copyright 2010 JP Hogan
All Rights Reserved

Home | Company Profile | Business Sectors | Client References | Projects | Gallery | Contact

http://www.888jphogan.com/ 2/26/2011

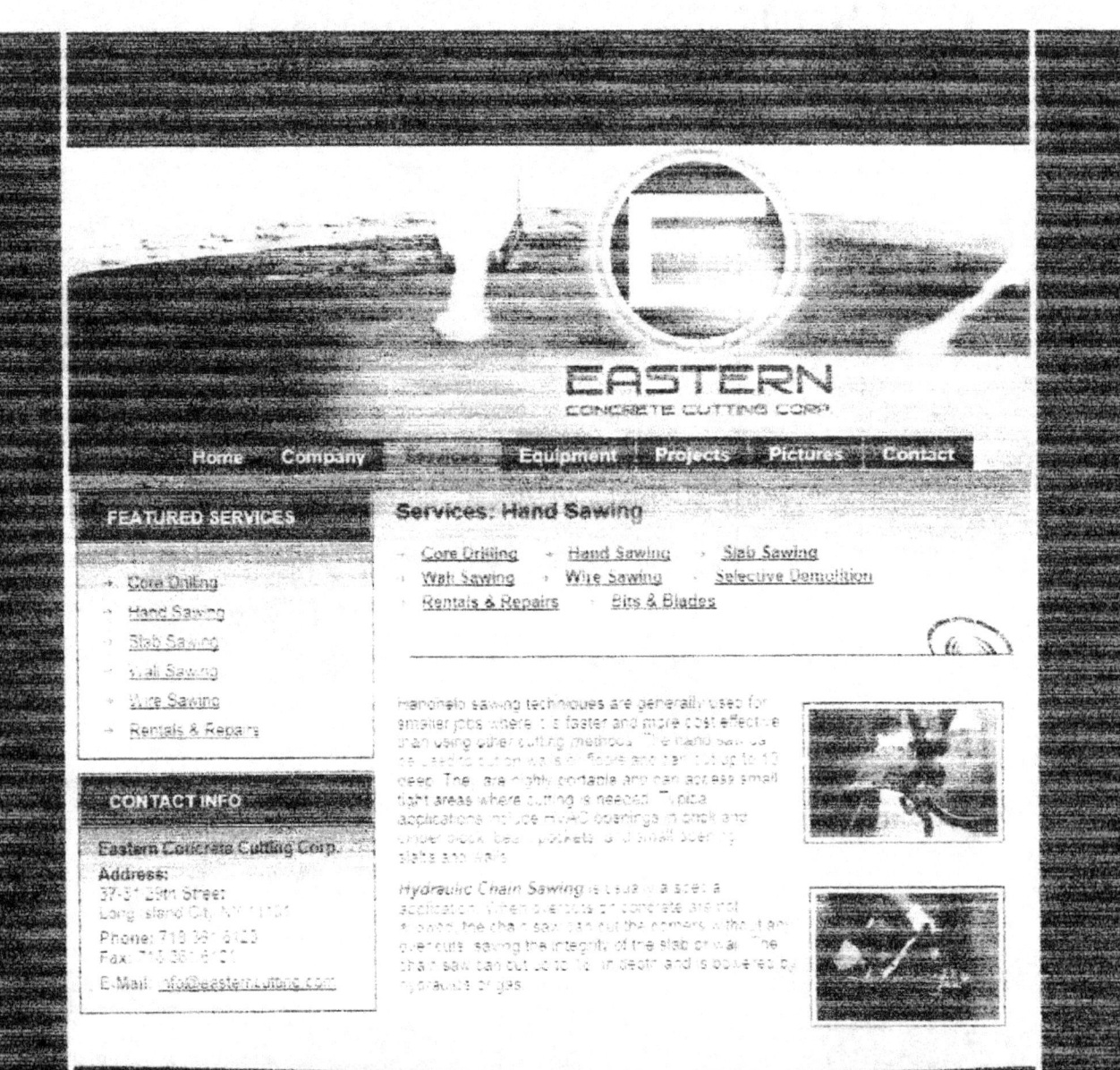

Exhibit "G"
1998 Consent

Schedules A and B

Schedule A

N/A

Schedule B

Prior to the commencement of the construction herein authorized, the Grantee shall submit to the Department of Environmental Protection, for review and approval, engineers scale drawings showing the proposed water main work and relocation of the manhole.

Prior to the opening of the museum, the Grantee shall submit to the Fire Department, for approval, a fire protection plan.

The Grantee shall submit plans of the construction herein authorized to the Art Commission for approval in accordance with the provisions of the New York City Charter and said construction shall not be commenced until such approval has been granted and a copy thereof filed with the Division of Franchises, Concessions and Consents of the Department of Transportation.

The Grantee shall file inspection reports with the Grantor at five-year intervals certifying the following:

a) The structural members were inspected by a professional engineer within the last six (6) months.

b) The load carrying capacity is sufficient to support the anticipated loading.

c) The non-load carrying members have been inspected and are secure.

The Grantee shall properly maintain all fire protection equipment and devices such as sprinkler systems, fire extinguishers, fire-proof self-closing doors, etc.

APPROVAL AS TO FORM OF A REVOCABLE CONSENT AGREEMENT BY STANDARD TYPE OF CLASS

AGENCY: Transportation

REVOCABLE CONSENT AGREEMENT: Owner

I hereby approve as to form the annexed revocable consent agreement by standard type of class. This approval is valid for one year and for a maximum of 300 consents.

The above approval is made on the express understanding that the substantive language of the subject revocable consent agreements will not be altered or changed in any way without prior submission to the office of the Corporation Counsel for approval provided, however, that blank spaces in the revocable consent agreements requiring names, dates, locations, dollar amounts or other similar details may be completed.

ACTING CORPORATION COUNSEL

Revised 6/15/2006

Note: DOT wasn't permitted to make any changes when renewing the consent in 2008

Currat, defective Consent

Schedule

The Grantee is required to have the tunnel inspected by qualified inspection personnel in accordance with the provisions of the New York State Department of Transportation Bridge Inspection Manual and file inspection reports with the Grantor at two-year intervals certifying the following:

a) The structural members were inspected by a professional engineer within the last six (6) months.

b) The load carrying capacity is sufficient to support the anticipated loading.

c) The non-load carrying members have been inspected and are secure.

The Grantee shall properly maintain all fire protection equipment and devices such as sprinkler systems, fire extinguishers, and fire-proof, self-closing doors etc.

DOT Staff may regularly inspect and photograph consented Structure to ensure that it is maintained properly.

Revised 6/03/2008
H D

Exhibit "H"

NEW YORK CITY Department of Transportation

JANETTE SADIK-KHAN, Commissioner

July 22, 2009

Mr. Robert Diamond
Brooklyn Historic Railway Association
599 East 7th Street, apt. 5A
Brooklyn, NY 11218

Re: Operations at site of revocable consent structure

Dear Mr. Diamond:

We have been made aware of the measures you take when conducting tours of the railroad tunnel that is the subject of the revocable consent entered into between the City of New York and the Brooklyn Historic Railway Association on July 1, 2009. In order to conduct any such further tours it will be necessary for a licensed engineer to draw up and submit to us for approval a certified maintenance and protection of traffic plan that will be implemented during any such scheduled tour. In order to assist you, we have put together a preliminary plan to be used as a guide for creating your plan. Please be advised that the plan that you submit to us will have to comply with the Manual of Uniform Traffic Control Devices.

Once approved by us, the plan will be incorporated into your revocable consent agreement and therefore all provisions of your revocable consent agreement shall apply to the plan, including but not limited to, the indemnity and hold harmless provisions, and insurance. No further tours may be conducted until this plan has been approved.

Please be further advised that you are not permitted to charge those who take your tours but you may accept voluntary donations.

Sincerely,

Emma Berenblit
Director of Revocable Consents

Encl.
EB:rs

NYC Department of Transportation
Office of Franchises, Concessions and Consents
55 Water Street, 9th Floor West, New York, NY 10041
T. 212.839.6551 F. 212.839.9695
www.nyc.gov/dot

 **Department of Transportation**

JANETTE SADIK-KHAN, Commissioner

July 22, 2009

Mr. Robert Diamond
Brooklyn Historic Railway Association
599 East 7th Street, apt. 5A
Brooklyn, NY 11218

Re: Operations at site of revocable consent structure

Dear Mr. Diamond:

We have been made aware of the measures you take when conducting tours of the railroad tunnel that is the subject of the revocable consent entered into between the City of New York and the Brooklyn Historic Railway Association on July 1, 2009. In order to conduct any such further tours it will be necessary for a licensed engineer to draw up and submit to us for approval a certified maintenance and protection of traffic plan that will be implemented during any such scheduled tour. In order to assist you, we have put together a preliminary plan to be used as a guide for creating your plan. Please be advised that the plan that you submit to us will have to comply with the Manual of Uniform Traffic Control Devices.

Once approved by us, the plan will be incorporated into your revocable consent agreement and therefore all provisions of your revocable consent agreement shall apply to the plan, including but not limited to, the indemnity and hold harmless provisions, and insurance. No further tours may be conducted until this plan has been approved.

Please be further advised that you are not permitted to charge those who take your tours but you may accept voluntary donations.

Sincerely,

Emma Berenblit
Director of Revocable Consents

Encl.
EB:rs

NYC Department of Transportation
Office of Franchises, Concessions and Consents
55 Water Street, 9th Floor West, New York, NY 10041
T: 212.839.6551 F: 212.839.9895
www.nyc.gov/dot

R.P. No: 1289

THE CITY OF NEW YORK
DEPARTMENT OF TRANSPORTATION
55 Water Street
New York, New York 10041

REVOCABLE CONSENT AGREEMENT (Owner)
(Modification)

WHEREAS The New York City Department of Transportation (the "Grantor"), by a revocable consent agreement dated July 1, 2008 (the "Consent") granted the Brooklyn Historic Railway Association, having its principal place of business at 599 East 7th Street, Brooklyn, NY 11218 (the "Grantee") consent to use and maintain a railroad tunnel, together with two public entrances, a manhole and ventilators, (the "Structure") in Atlantic Avenue from east of Columbia Street to west of Boerum Place, in the Borough of Brooklyn. The Consent will expire by limitation on June 30, 2018; and

IT IS HEREBY AGREED:

1. The Consent is hereby modified as follows:

(a) The Consent is amended to include the attached Maintenance and Protection of Traffic Plan submitted by Grantee (the "MPT Plan"). The Grantee shall comply with all provisions of the MPT Plan every time that Grantee accesses the Structure through the entrance on the roadway. Should Grantee fail to do so, Grantor may exercise all rights available to it, including its right to terminate the Consent. Grantee shall be solely responsible for all costs and expenses related to its operations, including implementation of the MPT Plan.

(2) The terms of this amendment shall be effective on the date of execution of this amendment and shall continue in full force and effect for the life of the Consent.

(3) Except as modified herein, the terms and conditions of the Consent shall remain in full force and effect throughout the term of the Consent.

(4) All provisions of the Consent shall apply to the MPT Plan, including but not limited to, the indemnity and hold harmless provision, and the insurance provisions.

2009-031222

In Witness Whereof, the parties hereunder have caused this amendment to a revocable consent to be executed.

GRANTOR:
NYC DEPARTMENT OF TRANSPORTATION
DIVISION OF FRANCHISES, CONCESSIONS & CONSENTS

By: _____
Anne Koenig
Executive Director

Accepted and agreed to:
GRANTEE:

By: _____Robert Diamond_____
(Signature)

_____Robert Diamond_____
(Print Name of Signatory)

_____President_____
(Title)

_____August 25 2009_____
(Date)

The foregoing amendment is hereby approved.
MICHAEL R. BLOOMBERG, MAYOR

By: _____
David Taylor-Fink, Associate Director for Program Administration
Mayor's Office of Contract Services

Dated, New York __9/8/__, 2009

Approved as to form
    Certified as to legal authority.

By: _____Sharon Cantor_____
Acting Corporation Counsel

_____AUG 2 1 2009_____
(Date)

2

Acknowledgment by Executive Director

State, City and County of New York, ss.:

On the 4 day of September, in the year 2009, before me, the undersigned, personally appeared Anne Koenig, personally known to me or proved to me on the basis of satisfactory evidence to be the individual(s) whose name(s) is(are) subscribed to the within instrument and acknowledged to me that he/she/they executed the same in his/her/their capacity(ies), and that by his/her/their signature(s) on the instrument, the individual(s), or the person upon behalf of whom the individual(s) acted, executed the instrument.

_____
Notary Public or Commissioner of Deeds

ROXANNE GAIR
Commissioner of Deeds
City of New York No. 2-10225
Commission Expires 2-1-70

Acknowledgment by Corporation

State, City, and County of New York, ss.,

On the 25th day of August, in the year 2009, before me, the undersigned, personally appeared Robert Diamond, personally known to me or proved to me on the basis of satisfactory evidence to be the individual(s) whose name(s) is(are) subscribed to the within instrument and acknowledged to me that he/she/they executed the same in his/her/their capacity(ies), and that by his/her/their signature(s) on the instrument, the individual(s), or the person upon behalf of whom the individual(s) acted, executed the instrument.

Velma Lewis

Notary Public or Commissioner of Deeds

VELMA P. LEWIS
Commissioner of Deeds
City of New York No. 4-5169
Certificate Filed in New York County
Commission Expires Feb. 1, 20__

PROJECT NO. BA-09-065

HISTORIC RAILWAY TUNNEL ACCESS, ATLANTIC AVENUE & COURT STREET, BROOKLYN, NY

MAINTENANCE AND PROTECTION OF TRAFFIC AND PEDESTRIANS PLANS

CLIENT: **BROOKLYN HISTORIC RAILWAY ASSOCIATION**
599 EAST 7TH STREET, STE.5A
BROOKLYN NY 11218

TRAFFIC CONSULTANT: **SAM SCHWARTZ ENGINEERING, PLLC**
611 BROADWAY, SUITE 415
NEW YORK, NY 10012

LOCATION PLAN

SITE PLAN

INDEX OF DRAWINGS

DWG. NO.	TITLE
G1	LOCATION PLAN/SITE PLAN
G2	INDEX OF DRAWINGS
G3	PHOTOS
MPT-1	MAINTENANCE AND PROTECTION OF TRAFFIC GENERAL NOTES
MPT-2	EXISTING CONDITIONS
MPT-3	EXISTING CONDITIONS
MPT-4	LEFT LANE CLOSURE ON ATLANTIC AVE EB
MPT-5	LEFT LANE CLOSURE ON ATLANTIC AVE EB
MPT-6	TRAFFIC CHANNELIZING DEVICES DETAILS (TYPE III BARRICADE AND CONE)

ATLANTIC AVENUE EB AT COURT ST, LOOKING WEST

ATLANTIC AVENUE EB AT COURT ST, LOOKING WEST

ATLANTIC AVE & COURT ST INTERSECTION, LOOKING NORTH

TUNNEL ENTRANCE MANHOLE

MAINTENANCE & PROTECTION OF TRAFFIC NOTES:

1. ALL MAINTENANCE & PROTECTION OF TRAFFIC WORK SHALL CONFORM TO THE NATIONAL MANUAL ON UNIFORM TRAFFIC CONTROL DEVICES FOR STREETS AND HIGHWAYS (MUTCD) AND THE NEW YORK STATE SUPPLEMENT TO THE NATIONAL MANUAL ON UNIFORM TRAFFIC CONTROL DEVICES FOR STREETS AND HIGHWAYS (NEW YORK STATE SUPPLEMENT) HEREBY DEFINED AND REFERENCED AS MUTCD. THE N.Y.S.D.O.T. STANDARD SPECIFICATION, SECTION 619 AS REVISED THROUGH 09-04-28, IS THE LATEST VERSION, EXCEPT AS MODIFIED IN THE PLANS AND PROPOSAL.

2. THE CONTRACTOR SHALL NOTIFY THE NEW YORK CITY POLICE DEPT. TRAFFIC DIVISION AT (212) 239-2500, THE NEW YORK CITY FIRE DEPT. OPERATIONS DIVISION AT (718) 999-7900, THE N.Y.D.O.T OFFICE OF CONSTRUCTION MITIGATION AT (212) 442-6775 AT LEAST TWO WEEKS IN ADVANCE OF THE COMMENCEMENT OF WORK ON A TRAVEL LANE OR SHOULDER MODIFICATION. SHALL BE IN WRITING AFTER RECEIPT OF CONCURRENCE FROM THE ENGINEER.

3. ALL CLOSURES SHALL BE COORDINATED WITH THE ENGINEER. A SCHEDULE OF WORK SHALL BE SUBMITTED AT LEAST TWO WEEKS IN ADVANCE FOR APPROVAL BY THE ENGINEER.

4. IF AT ANY TIME, IN THE OPINION OF THE ENGINEER, CONDITIONS SHALL WARRANT MODIFICATIONS TO THE SCHEMES SHOWN ON THIS OR OTHER MAINTENANCE & PROTECTION OF TRAFFIC DRAWINGS, THE CONTRACTOR SHALL PERFORM SUCH MODIFICATIONS, INCLUDING THE RE-OPENING OF ANY LANE CLOSURES ON AN EMERGENCY BASIS, A.O.B.E.

5. ALL CONSTRUCTION SIGNS SHALL MEET REQUIREMENT OF SECTION 619-2.02M HAVE AN ORANGE BACKGROUND, REFLECTORIZED, LEGEND ONLY AT NIGHT, WITH BLACK LETTERS AND BORDERS. 150mm ON LOCAL STREET SIGNS, UNLESS OTHERWISE DIRECTED BY THE E.I.C.

6. CONSTRUCTION SIGNS SHALL BE VISIBLE ONLY WHEN THE WORK THEY PERTAIN TO IS IN PROGRESS. CONSTRUCTION SIGNS HAVING CENTER HINGED SIGN PANELS OR FOLDING PORTABLE SIGN SUPPORTS SHALL BE FOLDED DOWN WHEN THE WORK THEY PERTAIN TO IS NOT IN PROGRESS. OTHER CONSTRUCTION SIGNS SHALL EITHER BE TAKEN DOWN OR BE REMOVED TO THE REQUIREMENTS OF SECTIONS 619-2.02M AND 645-3.09 OF THE N.Y.S.D.O.T STANDARD SPECIFICATIONS.

7. THE BOTTOM OF TEMPORARY CONSTRUCTION SIGNS SHALL BE A MINIMUM OF 2.1m ABOVE THE TRAVEL PAVEMENT AND THE EDGE SHALL OFFSET A MINIMUM OF 6m CLEAR OF THE TRAVEL LANE. AS SHOWN ON THE CONTRACT DOCUMENTS OR A.O.B.E. IF THE EQUIVALENT RECTANGULAR SIGN IF THE 6m CLEARANCE CANNOT BE MET USING RECTANGULAR SIGN, THE SIGN SHALL BE MOUNTED A MINIMUM OF 4m ABOVE THE TRAVEL PAVEMENT. UNDER NO CIRCUMSTANCES WILL THE CLIPPING OF SIGNS OR USAGE OF PREVIOUSLY CLIPPED SIGNS BE ALLOWED.

8. WARNING SIGNS SHALL BE LOCATED SO AS TO PROVIDE ADEQUATE VISIBILITY AND ADVANCE WARNING TO DRIVERS OF THE CONDITION AHEAD. AT A MINIMUM, FOR ANY PERMITTED INSTANCE OF OTHER SIGNS AND/OR TRAFFIC CONTROL DEVICES, NO STATIONARY MOUNTING OF CONSTRUCTIONS SIGNS SHALL BE PERMITTED ON UTILITY POLES OR OTHER ROADSIDE ELEMENTS. FOR NIGHTTIME CONSTRUCTION, SIGN PLACEMENT SHALL ALSO CONSIDER GLARE FROM HIGH MAST LUMINAIRES BEHIND THE SIGN AND HEADLIGHT FROM PARKING. SIGNS SHALL BE ORIENTED AS NEARLY PERPENDICULAR TO THE DIRECTION OF TRAFFIC. WHERE THE SIGN CANNOT BE LOCATED SO AS TO BE VISIBLE UNDER HEADLIGHT ILLUMINATION, OTHER ILLUMINATION SHALL BE CONSIDERED TO ENHANCE VISIBILITY OR THE SIGN SHALL BE RELOCATED.

9. IN REFERENCE TO THE MUTCD, THE FOLLOWING STIPULATIONS APPLY UNLESS OTHERWISE SPECIFIED BY THE ENGINEER:

   A) WHERE SIGNS ARE SHOWN IN BOTH DIAMOND AND RECTANGULAR FORMAT, THE DIAMOND SHAPES SHALL BE PERMITTED, EXCEPT AS MODIFIED BY NOTE 6 ABOVE OR A.O.B.E.

   B) WHERE SIGNS ARE SHOWN IN ALTERNATE SIZES, THE LARGEST SIZE MUST BE USED, UNLESS OTHERWISE SHOWN ON THE PLANS OR A.O.B.E.

10. IT IS THE RESPONSIBILITY OF THE CONTRACTOR TO FURNISH, INSTALL, MAINTAIN, AND REMOVE ALL CONSTRUCTION SIGNS AND TRAFFIC CONTROL DEVICES.

11. EXISTING SIGNS OF ANY TYPE THAT ARE RENDERED INAPPLICABLE BY CONSTRUCTION ACTIVITIES SHALL BE COVERED FOR THE DURATION OF THE WORK IN PROGRESS.

12. TRAVEL LANES SHALL BE SWEPT CLEAN BY THE CONTRACTOR PRIOR TO OPENING TO TRAFFIC. ALL PAVEMENT MARKINGS SHALL BE MAINTAINED OR RESTORED AFTER COMPLETION OF WORK.

13. TO ENSURE A SAFE TRAFFIC FLOW AT ALL TIMES, STORAGE OF MATERIALS AND EQUIPMENT, INCLUDING EMPLOYEES CARS, SHALL NOT BE PERMITTED WITHIN THE ROADWAY. STORAGE OF OTHER SIGNS AND TRAFFIC AREAS SHALL BE SEPARATED FROM THE TRAVELED WAY BY A CLEAR SPACE OF 3m MINIMUM WIDTH, UNLESS SUCH STORAGE IS PLACED BEHIND TEMPORARY CONCRETE BARRIER OR OTHER PERMANENT ROADSIDE BARRIER INSTALLATION.

14. THE MAINTENANCE & PROTECTION OF TRAFFIC SCHEMES SHOWN ON THE PLANS OR PROPOSAL ARE TO PROTECT THE TRAVELING PUBLIC. IT IS THE RESPONSIBILITY OF THE CONTRACTOR TO PROTECT THE WORKERS AND PROVIDE THEM WITH SAFE ACCESS TO WORK SITES.

15. LANE CLOSURES SHALL BE POSITIONED TO PROVIDE OPTIMUM VISIBILITY, I.E., BEFORE CURVES AND CRESTS, AND SHOULD BE LOCATED AWAY FROM OTHER CONFLICT POINTS, SUCH AS ON-RAMPS AND INTERSECTIONS. WHENEVER POSSIBLE, LANE CLOSURES SHALL ALSO BE LOCATED SO AS TO AVOID BRIDGES AND UNDERPASSES AND OTHER LOCATIONS WITHOUT AVAILABLE ESCAPE PATHS.

16. WHEN WORKING ON THE MEDIAN WITH ONLY ONE SPEED LANE CLOSED, THE CONTRACTOR SHALL PLACE CONES ALONG THE GUTTER LINE OF THE OPENED SPEED LANE, SPACED AT 12m INTERVALS FOR THE ENTIRE LENGTH OF THE WORK ZONE.

17. THE CONCURRENT CLOSURES OF LEFT AND RIGHT LANES SHALL NOT BE PERMITTED WITHIN A 3200m DISTANCE BETWEEN CONSECUTIVE CLOSURES IN THE SAME TRAVEL DIRECTION, EXCEPT AS MODIFIED BY SPECIAL NOTES IN THE PROPOSAL.

18. THE CONTRACTOR SHALL PROVIDE A FLAGGER WITH APPROPRIATE SIGNING WHENEVER OPERATIONS INTERFERE WITH TRAFFIC. EXAMPLES INCLUDE, BUT ARE NOT LIMITED TO, DELIVERY/REMOVAL OF MATERIALS OR EQUIPMENT, LIFTING OPERATIONS, AND ANY OTHER ACTIVITIES SO DESIGNATED BY THE ENGINEER. ALL FLAGGERS USED MUST BE FORMALLY TRAINED IN PROPER FLAGGING PROCEDURES AND TMBA CERTIFIED.

19. THE SOLE DUTY OF THE FLAGGER SHALL BE TO DIRECT TRAFFIC PROPERLY AT ALL TIMES. FLAGGERS SHALL NOT BE USED TO MOVE TEMPORARY SIGNS OR ASSIST IN OTHER WORK. IN ACCORDANCE WITH THE MUTCD FOR HAND SIGNALING DEVICES AND SIGNALING PROCEDURES SHALL BE MET. STOP/SLOW PADDLES SHALL BE USED RATHER THAN FLAGS. FOR MOST LONG-TERM FLAGGING OPERATIONS, CLEAN, WELL-FITTING ORANGE VESTS AND HARD-HATS SHALL BE WORN BY ALL FLAGGERS. FLAGGING STATIONS SHALL BE LOCATED TO PROVIDE ADEQUATE SIGHT DISTANCE AND VISIBILITY TO THE EXTENT POSSIBLE, AN ADEQUATE ESCAPE PATH SHALL BE PROVIDED FOR THE SAFETY OF THE FLAGGER AND THE SAFE RECOVERY OF VEHICLES THAT FAIL TO RESPOND TO FLAGGER DIRECTIONS. ANY FLAGGING DEFICIENCIES SHALL BE CORRECTED IMMEDIATELY, OR THE CONTRACTOR SHALL BE REQUIRED TO CEASE OPERATIONS UNTIL A COMPETENT FLAGGER IS OBTAINED.

20. FOR LONG-TERM OR OVERNIGHT OPERATIONS, THE CONTRACTOR SHALL SUBSTITUTE DRUMS FOR CONES, THE FIRST TWO CHANNELIZING DEVICES AT THE BEGINNING OF LANE CLOSURES SHALL BE FITTED WITH WARNING LIGHTS IN ACCORDANCE WITH MUTCD IN ADDITION, THE CONTRACTOR SHALL OBSERVE THE FOLLOWING:

    - TYPE A (LOW INTENSITY) OR TYPE B (HIGH INTENSITY) FLASHING WARNING LIGHTS ARE TO BE USED AS THE FIRST TWO LIGHTS IN A LONGITUDINAL SERIES AND FOR MARKING ISOLATED HAZARDS.

    - TYPE C (LOW INTENSITY) STEADY BURNING WARNING LIGHTS MAY ONLY BE USED ON TEMPORARY CONCRETE BARRIER OR OTHER NON-REFLECTIVE FEATURES LOCATED CLOSE TO TRAVEL LANES. THEY ARE NOT TO BE USED ON REFLECTORIZED CHANNELIZING DEVICES UNLESS JUSTIFIED BY REDUCED VISIBILITY OR HEAVY FOG.

    - TYPE A AND C SHALL BE USED FOR NIGHTTIME APPLICATIONS. TYPE B SHALL BE USED FOR DAYTIME APPLICATIONS, AND MAY BE USED AT NIGHT WHERE THE CHANNELIZING DEVICES ARE LEFT IN PLACE DURING THE DAY AND WHERE ADDITIONAL EMPHASIS IS DESIRABLE.

LEGEND

- CONSTRUCTION SIGNS
- WORK AREA
- SHADOW VEHICLE WITH TRUCK MOUNTED ATTENUATOR AND A FLASHING ARROW PANEL

MODES OF FLASHING ARROW PANEL:
- LEFT
- RIGHT
- DUAL
- CAUTION

- PORTABLE FLASHING ARROW BOARD, USED AS INDICATED ABOVE
- 900mm DRUM, 450mm MINIMUM DIAMETER
- VEHICLE ARRESTING BARRIER
- FLAGGER
- TYPE III BARRICADE WITH WARNING LIGHTS
- WARNING LIGHT AND TYPE
- TEMPORARY CONCRETE BARRIER
- TEMPORARY SAND BARREL ARRAYS
- PORTABLE QUADRILATERAL TERMINAL IMPACT ATTENUATOR

ABBREVIATIONS

- A.O.B.E. — AS ORDERED BY ENGINEER
- E.I.C. — ENGINEER IN CHARGE
- T.M.A. — TRUCK MOUNTED ATTENUATOR
- N.Y.S.D.O.T. — NEW YORK STATE DEPARTMENT OF TRANSPORTATION
- N.Y.C.D.O.T. — NEW YORK CITY DEPARTMENT OF TRANSPORTATION
- MUTCD — NATIONAL MANUAL ON UNIFORM TRAFFIC CONTROL DEVICES, NEW YORK STATE SUPPLEMENT TO THE NATIONAL MANUAL ON UNIFORM TRAFFIC CONTROL DEVICES

BROOKLYN HISTORIC RAILWAY ASSOCIATION
MAINTENANCE AND PROTECTION OF TRAFFIC
GENERAL MPT NOTES

MPT PLANS FOR RAILROAD TUNNEL ACCESS AT ATLANTIC AVE. & COURT ST. BROOKLYN, NY

SSE PROJECT #: BA-09-065
CONTRACT NO.: BA-09-065
SHEET NO.: 4 / TOTAL SHEETS: 9

DRAWING NO.: MPT-1
SCALE: N/A
DATE: 07-28-2009
SAM SCHWARTZ ENGINEERING, NEW YORK, NY

DESIGN SUPERVISOR: A. SIDDIQUI
JOB MANAGER: A. ZAFAR
DESIGNED BY: A. ZAFAR
CHECKED BY: A. SIDDIQUI
ESTIMATED BY: A. ZAFAR
DRAFTED BY: A. NULLA
CHECKED BY: A. ZAFAR
CADD REG: 11

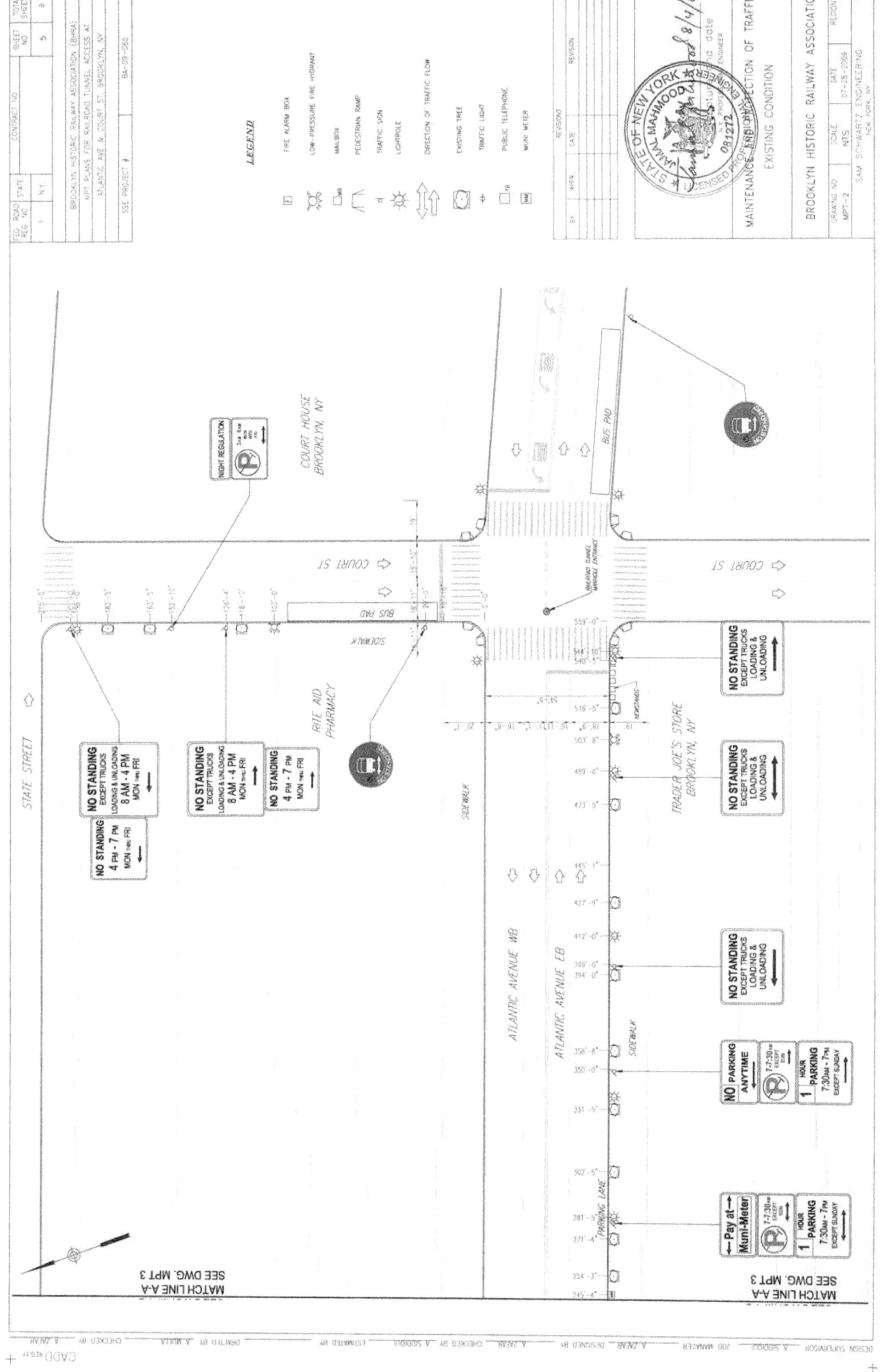

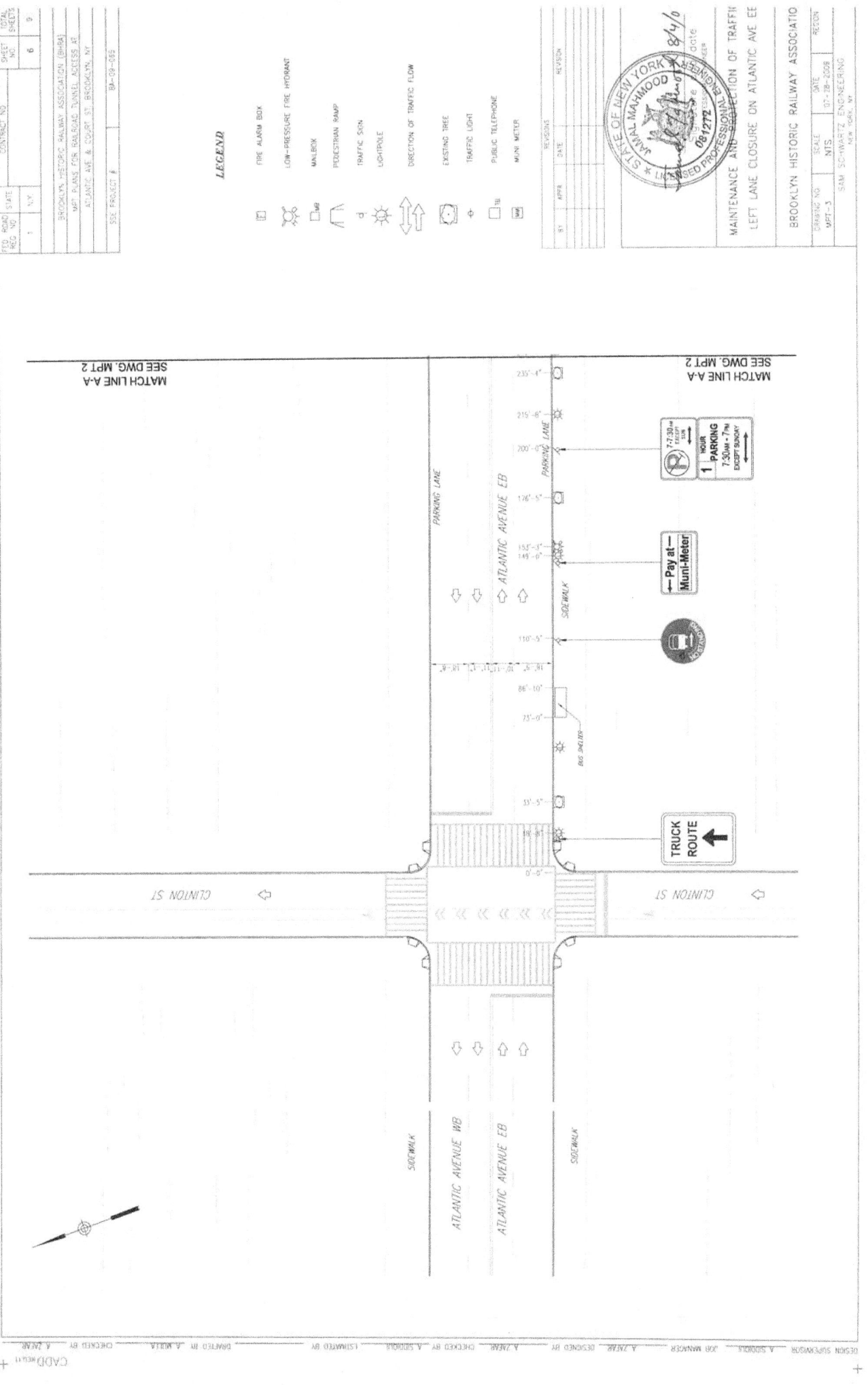

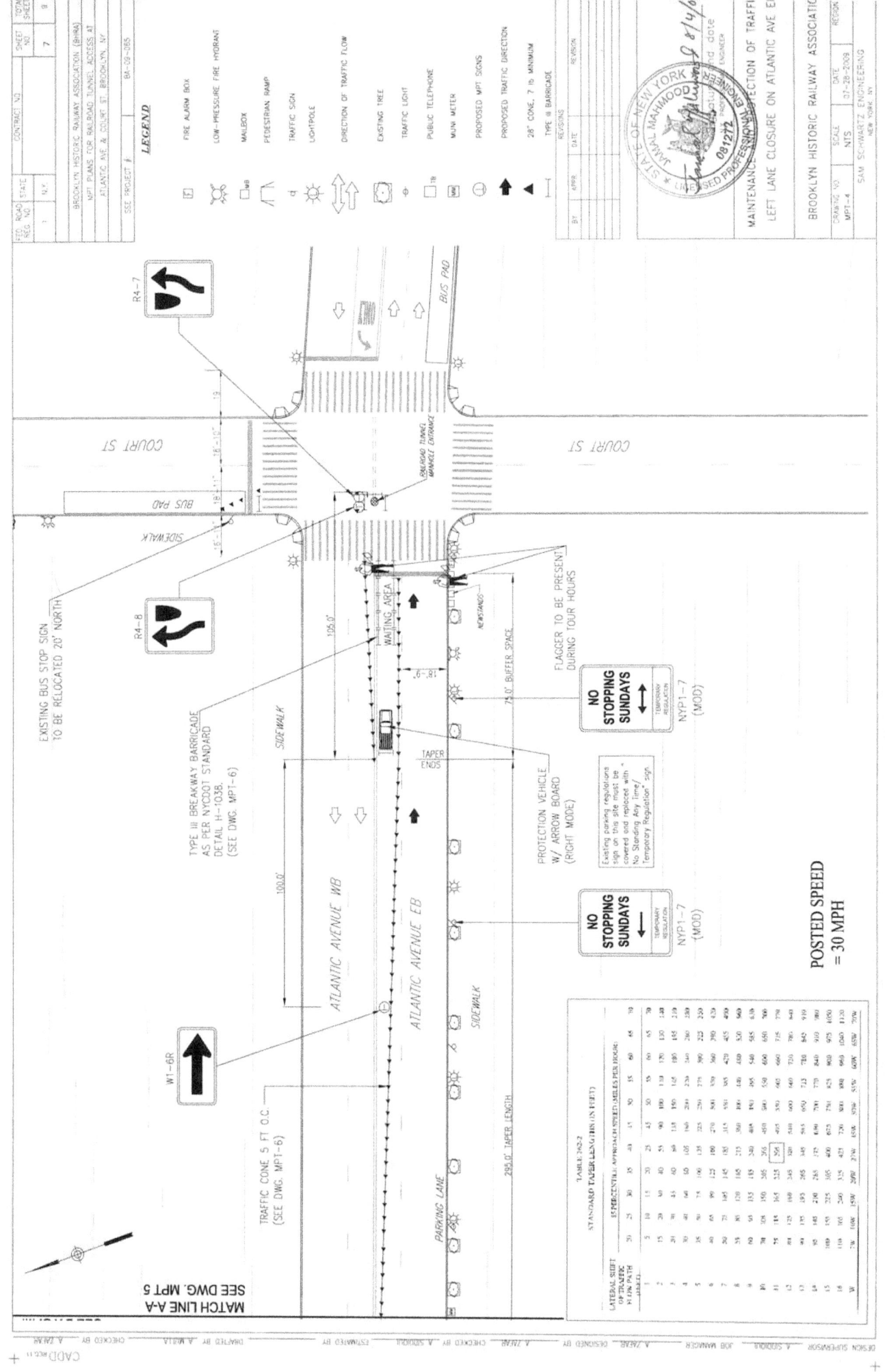

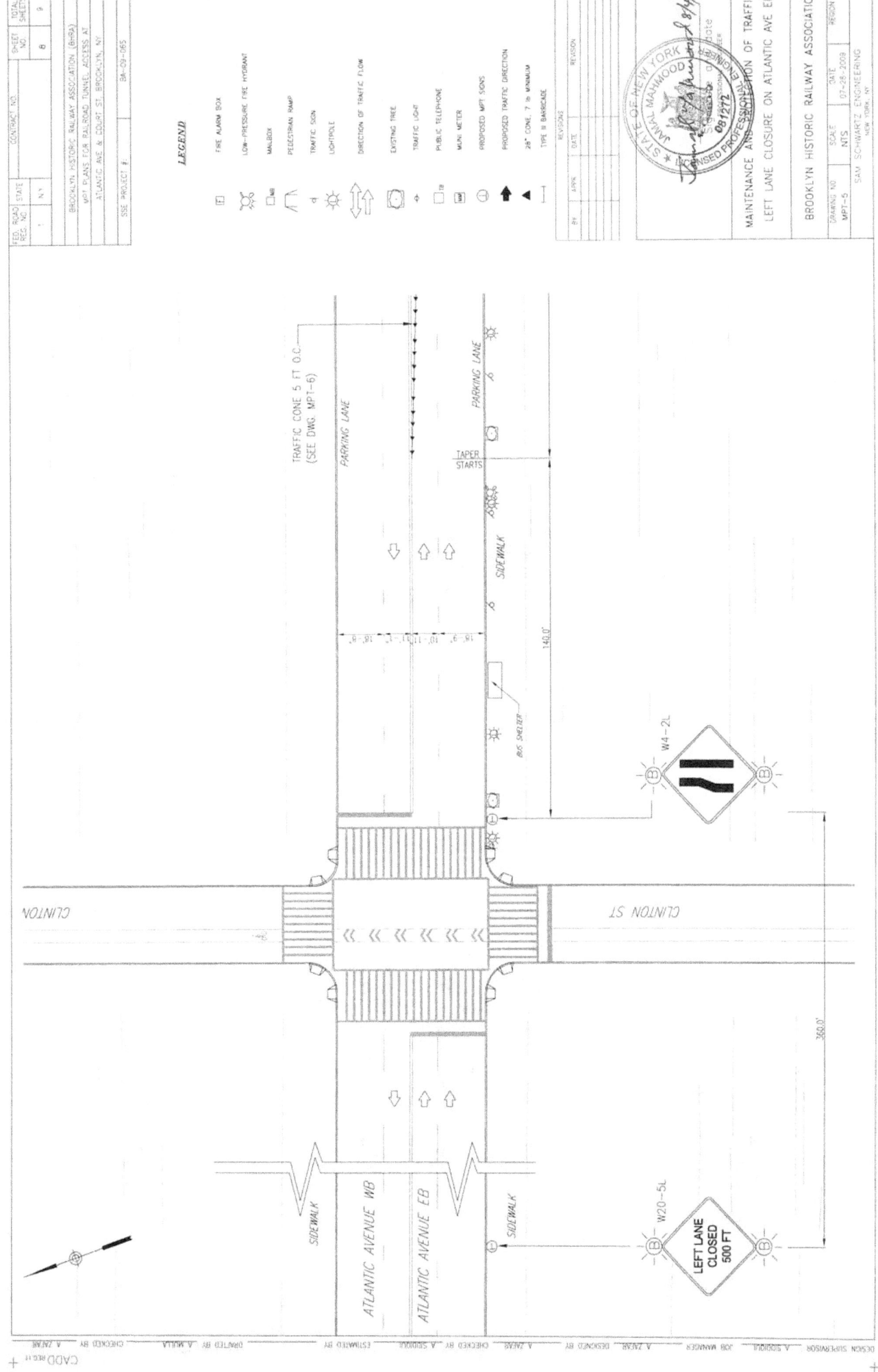

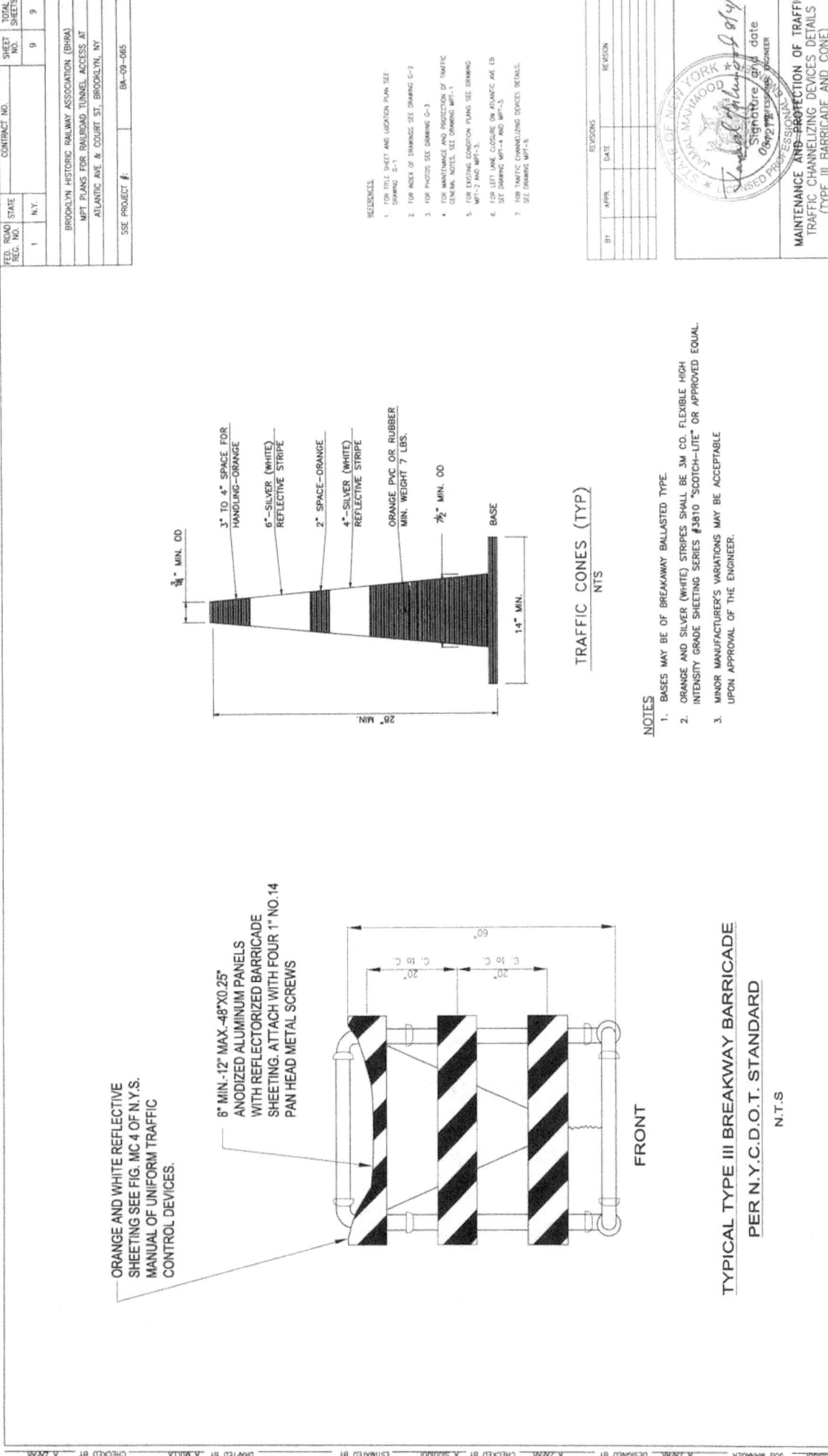

LMW Engineering Group, LLC

2539 Brunswick Ave. Linden, NJ, 07036   Tel.(908) 862-7600 Fax(908) 862-8998   WWW.LMW-ENG.COM

March 25, 2009

BROOKLYN HISTORIC RAILWAY ASSOCIATION
599 E 7th Street, Suite 5A
Brooklyn, NY 11218

ATTN.:   Robert Diamond, President

RE: Atlantic Avenue Tunnel,
        between Boerum Place and Columbia Street

SUBJECT: <u>Structural Integrity Evaluation</u>

Dear Mr. Diamond;

Pursuant to your request, we have performed an evaluation of the structural integrity of the referenced tunnel structure, based on visual inspection tour of the tunnel site, and review of reports of previews evaluations performed by others. This letter-report presents the results of our evaluation.

The inspection tour of the tunnel was performed by our senior engineer, Mr. Tony Onyeagoro, P.E., in the afternoon hours of Wednesday, March 25, 2009, assisted by one of your associates. Access and egress to the site were through a manhole located at the middle of the roadway, intersection of Atlantic Avenue and Court Street, Borough of Brooklyn. Visibility in the tunnel was generally subdued, but details of the structural elements were clearly observable using a flash light. Select, typical portions of the structure details were captured using flash-enabled *Canon PowerShot A540* camera.

The tunnel structure consists of a masonry block arched dome upper section, supported on either side by a wall made of varying-size stone or rock quarry, embedded in a matrix of very strong grout mix. The composition of the structural elements is relatively consistent for the entire length of the tunnel. Evidence of removed rail lines and ballasts are observable through out the general grade of the tunnel, which is firm with no sign of moisture or subsidence. The general condition of the tunnel structural elements, relative to previous reports, is impressively devoid of any sign of deterioration. All masonry block units appear to be intact, with no visible loosening of joints. Similarly, there was no observable loss of filler materials within the walls' stone matrices.

Accordingly, in concurrence with the findings of previous evaluations, please be advised that:

1. The tunnel structural members were inspected by our engineer.
2. The load carrying capacity of the tunnel is sufficient to support the anticipated loading (overburden and live load).
3. The non-load carrying members are secure.
4. The structural integrity of the tunnel is sound and has not been compromised by aging.

Certified Minority Business Enterprise NY/NJ
Engineering Consultation ● Design ● Inspection ● Testing

5. The tunnel can be considered safe under its current use for visitors and tourist attraction.
6. There is no evidence that any form of maintenance or repair work is necessary at this stage.
7. The tunnel can be safely, with relatively minimum rehabilitation effort, mostly esthetic, be utilized as a museum or similar facility.
8. In summary, the tunnel, as inspected by us, is a safe and sound structure.

We appreciate the opportunity to provide this service for you, and look forward to continuing involvement in this extraordinarily noble venture. If you have any questions regarding this letter-report, please feel free to call us.

Very truly yours,
LMW ENGINEERING GROUP, LLC

Tony C. Onyeagoro, P.E.
Senior Project Engineer
NYS P.E. License # 063593

Exhibit "J"

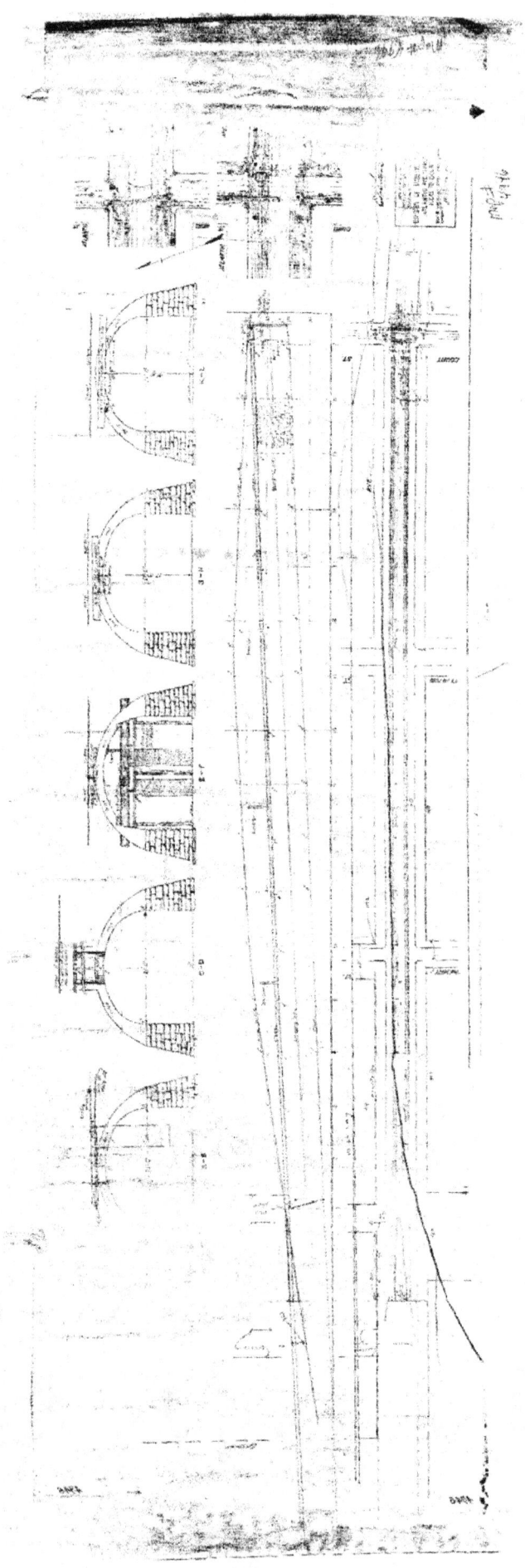

Subject: RE: Completed Atlantic Avenue Tunnel Report
From: Berenblit, Emma (eberenblit@dot.nyc.gov)
To: rdiamond@brooklynrail.net;
Cc: akoenig@dot.nyc.gov; djehn@dot.nyc.gov;
Date: Friday, March 27, 2009 10:58 AM

Dear Robert

Thank you for sending us completed inspection report for Atlantic Avenue tunnel. Inspection report shows that the tunnel is in good condition and safe. You may continue usage of the tunnel. However you need to comply with all applicable laws, rules and regulations when conducting operations including obtaining any and all permits required by the City of New York. If you close streets in order to access the tunnel, you need to obtain the requisite street closing permits from DOT.

From: Robert Diamond [mailto:rdiamond@brooklynrail.net]
Sent: Thursday, March 26, 2009 7:39 PM
To: Berenblit, Emma
Subject: Completed Atlantic Avenue Tunnel Report

Dear Emma, please find attached the completed preliminary structural report on the Atlantic Avenue tunnel. According to the report, this is a safe and sound structure, please give special attention to items 5 and 6 in the report. Also attached are the photos.

I assume we can now proceed with our usual access to the tunnel.

Thanks again for your assistance and understanding in this matter.

Bob Diamond

****

```
This message and any attachments are solely for the individual(s) named above and others who have been specifically authorized to
receive such and may contain information which is confidential, privileged or exempt from disclosure under applicable law. If you
are not the intended recipient, any disclosure, copying, use or distribution of the information included in this message and any
attachments is strictly prohibited. If you have received this communication in error, please notify us by reply e-mail and
immediately and permanently delete this message and any attachments.

Thank you.


NYC - Department of Transportation
```
****

NEW YORK CITY DOT Department of Transportation

JANETTE SADIK-KHAN, Commissioner

February 11, 2010

Mr. Robert Diamond
Brooklyn Historic Railway Association
599 East 7th Street, Apt. 5A
Brooklyn, NY 11218

Re: Schedule of Visits to Atlantic Avenue Railroad Tunnel

Dear Mr. Diamond:

This letter is in regard to the revocable consent authorizing the Brooklyn Historic Railway Association to maintain and use a railroad tunnel, together with two public entrances, a manhole and ventilators, in Atlantic Avenue from east of Columbia Street to west of Boerum Place, in the Borough of Brooklyn.

Please provide our office with a schedule of the dates and times of planned visits to the railroad tunnel over the next two months. The schedule should include all planned group and individual visits, whether for tours, maintenance and operation, or any other purpose.

Please submit the schedule within one week of the date of this letter. Thank you.

Sincerely,

Emma Berenblit
Director, Revocable Consents

cc: Anne Koenig
Franco Esposito

NYC Department of Transportation
Office of Franchises, Concessions and Consents
55 Water Street, 9th Floor SW, New York, NY 10041
T: 212.839.6550 F: 212.839.9895
www.nyc.gov/dot

LMW Engineering Group, LLC
2539 Brunswick Ave. Linden, NJ, 07036   Tel. (908) 862-7600 Fax (908) 862-8998
WWW.LMW-ENG.COM

February 17, 2012

Mr. Robert Diamond, President
Brooklyn Historic Railway Association
599 E 7th St.;
Suite 5A
Brooklyn, NY 11218

Reference: Historic Atlantic Ave Railway Tunnel

Dear Mr. Diamond:

As requested, I attended a meeting with city officials for the referenced site. It was my impression that I, as the design engineer of the proposed new egress access, was there to explain our design, so, the city could accept the proposed design and permit its construction for the expanded use of the Tunnel. However, focus of the meeting was actually all about various legal and administrative issues about the Tunnel itself. Following city officials were present during the meeting:

- Joseph Palmieri, Brooklyn Borough Commissioner (NYC DOT);
- Chief James Manahan, Assistant Chief of Operations (NYFD);
- Julian Bazel, Legal Counsel (FDNY);
- Warren Shaw, Counsel (NYC Department of Law).

It should be noted that I requested to access to the Tunnel as soon as possible, so any outstanding issues with the design can be addressed.

If we can be of any further assistance to you with regard to this matter, please feel free to call our office.

Very truly yours,
LMW ENGINEERING GROUP, LLC

Jieming Wang, PE

Engineering Consultation ● Design ● Inspection ● Testing

JIEMING WANG, P.E.

In his capacity as the managing partner and president of LMW Engineering Group, Jieming Wang oversees the day-to-day operations of the company, in addition to the to the overall corporate management and planning. Since starting the company in 1998, Jieming has supervised and/or participated in various capacities in structural design, value engineering, engineering consultation, site condition related engineering, structural monitoring and inspection, deep foundation related engineering on numerous governmental and private projects.

Prior to joining our management team, Jieming has spent over (8) eight years working as a geotechnical engineer, then a senior geotechnical engineer at William F. Loftus Associates. During that period, in addition to his regular engineering supervision duty, he was also appointed by TNO of Holland, as its North American representative. His responsibilities included promoting its FPDS Technology in the geotechnical field and providing technical support to its clients in North America. Through the involvement in promoting the FPDS testing equipment in the US and various instrumented testing projects, he gained a better understanding about the problems and difficulties in the deep foundation and geotechnical field. It enables him to provide better services to our clients.

Jieming Wang earned his BS from Beijing University of Science and Technology from China and received his MS and MBA from the University of Wollongong in Australia. Jieming is a licensed Professional Engineer in New York, New Jersey, Connecticut, and Florida.

Exhibit "I" 1 of 3

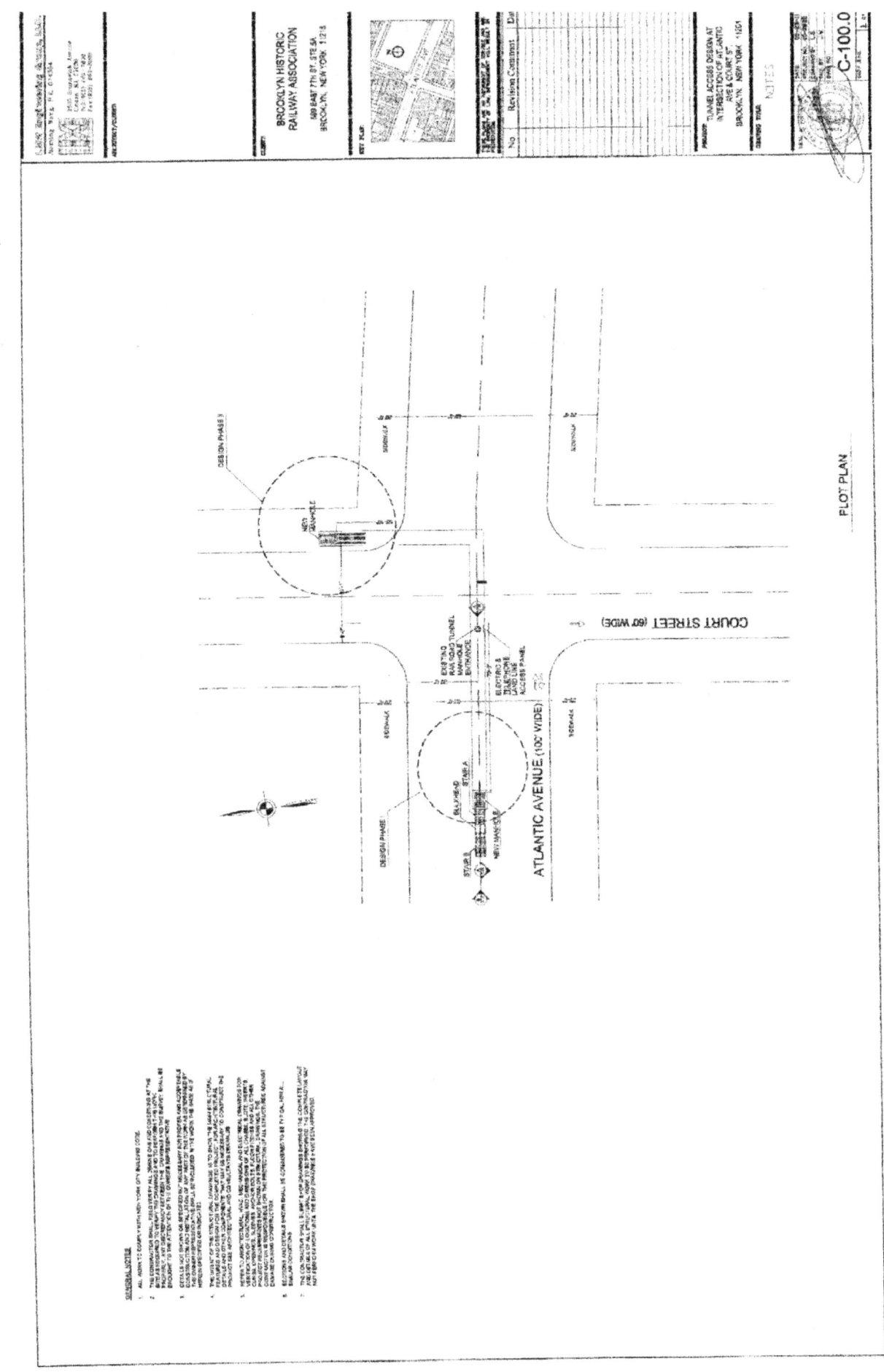

Exhibit "I" 2 of 3

Exhibit "I" 3 of 3

CALENDAR
of the
Board of Estimate

The City of New York

THURSDAY, OCTOBER 9, 1986

MEETING at 10:30 A.M.
in the
CITY HALL
Borough of Manhattan

(Volume No. 5)
Prepared by Miss Jacqueline Galory, Calendar Clerk, under the direction
of Theodore H. Meekins, Secretary, Board of Estimate

SAVE WATER

Thursday, October 9, 1986   152

by the Bureau of Highways. The proposed lampposts are identical in design to the lampposts now in Central Park and Battery Park. They will have a 175 watt metal halide lamp and will be operated by time clock from dusk to 1:00 a.m. The 8 existing city street lamps along 23rd Street between 5th and 6th Avenue provide ample illumination allowing the decorative lamps to be turned off at 1 a.m.

The 9 lampposts are 13 feet high and are placed along the street side of the sidewalk at an interval of approximately 22 feet. Their placement will not conflict with the existing street lights. Because the sidewalk is 17'5" wide, the posts will not interfere with pedestrian use of the sidewalk.

Community Board No. 5 voted in favor of the application.

At the public hearing held by the City Planning Commission on August 6, 1986, there were no speakers.

The City Planning Commission finds the proposal to be in the public interest and has approved the grant of a revocable consent.

NOTE: Pursuant to Section 197-c of the New York City Charter, the Board of Estimate shall hold a public hearing and take final action on this matter on or before November 1, 1986.

Close the hearing.

Resolution for adoption.

No. 47

R-5477, 5812

PUBLIC HEARING on the petition of the Brooklyn Historic Railway Association for consent to maintain and use an abandoned railroad tunnel in Atlantic Avenue from east of Columbia Street to west of Boerum Place, together with an existing manhole, and to construct, maintain and use entrances and ventilators, Borough of Brooklyn.

On September 25, 1986 (Cal. No. 187), the Board of Estimate set October 9, 1986 as the date for a public hearing on the petition.

NOTICE of this hearing has been duly advertised in The City Record, as required by law.

REPORT of the Director of Franchises, dated September 26, 1986, stating that the tunnel will be used as a museum and exhibition space.

The Department of Environmental Protection, Bureau of Water Supply and Bureau of Sewers, the New York City Department of Transportation, the Bureau of Electrical Control and the Fire Department have advised the Bureau of Franchises that they have no objections to the petition.

Thursday, October 9, 1986

The Bureau of Franchises has no objections to offer, and as the administrative departments of City government having jurisdiction find none, it is recommended that the consent be granted, only during the pleasure of the Board, revocable at any time, but in no event to extend beyond a term of ten years from the date of the approval of the consent by His Honor, the Mayor.

Compensation—Five percent of gross receipts, but not less than $2,000 per annum.

Security Deposit—$2,000.

Insurance—The grantee shall maintain insurance coverage in the minimum amount of $1,000,000 for bodily injury and property damage, until commencement of construction. Thereafter, in the minimum amounts of $5,000,000 for bodily injury, including death, and $1,000,000 for property damage.

REPORT of the City Planning Commission (C 850022 GFK), dated August 27, 1986, stating that the site is located between the Brooklyn Heights Historic District to the north, and the Cobble Hill Historic District to the south. Both communities are predominantly residential, with local commercial uses. The zoning of the site is R6 with C1-3 and C2-3 mapped within the R-6 District, and it is partly within the Special Atlantic Avenue District. Atlantic Avenue is low scale, predominantly commercial (restaurants, antique stores, local convenience shopping), along with some residential uses on the upper floors.

This railroad tunnel was completed in 1844, and is believed to be the first time that the cut and cover method of tunnel construction was used. Sixty years later the first New York City Subway Line was constructed in this way. The Long Island railroad was built on the surface of Atlantic Avenue, crossing through Queens and the then City of Brooklyn, until the tracks reached Boerum Place. Because of the relative steep grades and heavy surface traffic on Atlantic Avenue, the portion of the line from Boerum Place to Columbia Street was constructed under the street. The tunnel was used until it was abandoned and sealed up in 1861. In August, 1980, the current President of the Brooklyn Historic Railway Association rediscovered this tunnel. The abandoned tunnel is 2,517 feet in length, of which approximately 1,500 feet is open for use. The tunnel, of brick construction, is 21 feet in width and 17 feet in height.

The Atlantic Avenue tunnel would be operated by the Brooklyn Historic Railway Association as a non-profit museum, with exhibition space for transportation and historical subjects. An electric trolley would be used in the tunnel, moving passengers along tracks from Court Street to Hicks Street and back.

Entrance will be from a subway type kiosk on the northeast corner of Atlantic Avenue and Court Street. There will be an emergency exit at the southwest corner of Atlantic Avenue and Hicks Street. The tunnel will be vented mechanically through three capped air shafts located on Atlantic Avenue, which will lead to six air vents, three on the south side of Atlantic Avenue and three on the north side.

Thursday, October 9, 1986   154

The vents will be approximately twelve feet high and will be located in the sidewalk.

The rediscovery of the long abandoned Long Island Railroad tunnel under Atlantic Avenue is a significant event in the history of Brooklyn and of railroad technology. The Long Island Railroad was originally constructed to provide a part of a Boston to New York link, emanating from Boston by rail, crossing Long Island Sound by steamboat, and then traversing Long Island by the L.I.R.R. to its western tip at Columbia Street near the Brooklyn waterfront. From here passengers could take a ferry to South Ferry, in Manhattan. The tunnel was used by steam trains until 1861, when it was abandoned by the Railroad and sealed. In that year the L.I.R.R. moved its western terminus to Hunter's Point in Queens, where passengers could take a ferry to Manhattan. By that time the New York, New Haven and Hartford Railroad had been constructed and the Boston to New York link by way of Long Island was abandoned, with the Long Island Railroad becoming a commuter railroad for Long Island, which it remains today. The Atlantic Avenue tunnel was gradually forgotten, but became famous in local Brooklyn lore.

The application was reviewed by the Department of Environmental Protection and the Department of City Planning pursuant to the New York State Environmental Quality Review Act and the New York City Environmental Quality Review regulations. A Conditional Negative Declaration was issued on February 20, 1986. The Conditions require archaeological investigation of the site of the proposed project prior to commencement of any preconstruction or construction work taking place on the site.

Brooklyn Community Boards Nos. 2 and 6 and the Brooklyn Borough Board voted in favor of the application.

At the public hearing held by the City Planning Commission on August 6, 1986 two speakers, including the applicant, appeared in favor of the application. The speakers said this project was important to the history of transportation in Brooklyn and the economic development of downtown Brooklyn.

Downtown Brooklyn is currently experiencing an increase in development activity, including a number of large development projects such a Atlantic Terminal, Brooklyn Center, and Metrotech and others. This tunnel is located near these projects, and it is expected to contribute to the overall increase of the economic and cultural development of this area. The tunnel will also be used as a transportation museum. Exhibits will show the history of transportation and of the old City of Brooklyn. The Brooklyn Historic Railway Association will be placing an 1890's vintage electrically operated trolley in the tunnel, moving visitors past the educational exhibits as the trolley moves on tracks from one end of the tunnel to the other, and back again. This tunnel will be a unique educational center for the residents and visitors to New York City.

The funding of the tunnel opening, kiosk, and the exhibits is expected to be provided by the Atlantic Avenue Association Local Development Corporation. The

Thursday, October 9, 1986

applicant has indicated that once this revocable consent is granted, the applicant, the Brooklyn Historic Railway Association, will ask the Board of Estimate, Bureau of Franchises, to include the Atlantic Avenue Association Local Development Corporation on the revocable consent. The applicant will also be required by the Bureau of Franchises to post the appropriate insurance for liability against personal injury and property damage. The New York City Department of Transportation will require the applicant to post a bond so that if the proposed construction work is not complete, the sidewalk and streets will be restored to their prior condition, and the tunnel would be bulkheaded and backfilled, as it presently is.

The City Planning Commission has approved the grant of a revocable consent for a period of fifty years to use and maintain an abandoned railroad tunnel in Atlantic Avenue from east of Columbia Street to west of Boerum Place, including construction, operation and maintenance of necessary public entrances, ventilators and facilities for the accommodation of the public, for use as a museum and exhibition space.

NOTE: Pursuant to Section 197-c of the New York City Charter, the Board of Estimate shall hold a public hearing and take final action on this matter on or before November 1, 1986.

Close the hearing.
Resolution for adoption.

Matters Laid Over

No. 48

R-9318

IN THE MATTER OF a
COMMUNICATION dated May 16, 1986, from the Acting Corporation Counsel, transmitting resolution and Supplemental Agreement and stating that the Board at its meeting on September 18, 1980 (Cal. No. 53-B), adopted Map No. 11962 showing the elimination, discontinuing and closing of portions of Caroll and Schofield Streets and Fordham Place, in the Borough of The Bronx.

The Board at its meeting of May 26, 1983 (Cal. No. 511), accepted an agreement between City Island Boatyard Limited Partnership (the applicant) and the City of New York relating to said map change.

This office has been informed by the Topographical Bureau of the Office of the President, Borough of The Bronx, that said map contains technical and mathematical errors. Said bureau prepared a revised map correcting such errors and submitted same to this office for submission to the Board.

1968 BUILDING CODE
OF THE CITY OF NEW YORK

Plus Selected Rules of the Department of Buildings
LOCAL LAW NO. 76 Effective Dec. 6, 1968
INCLUDES AMENDMENTS To July 1, 2008

MICHAEL R. BLOOMBERG
MAYOR

ROBERT D. LiMANDRI
Commissioner

PREFACE

This revision brings the 1968 Building Code current to July 1, 2008.

When enacted by the City Council on October 22, 1968, the 1968 Building Code was hailed as a great improvement over the anachronistic 1938 Building Code and included what was then the latest thinking in building code science, incorporating advances in technology and construction that had been made following the Second World War. Over the years, the Council amended the 1968 Building Code to address certain changes as needed; however, the 1968 Building Code never enjoyed a complete overhaul, gradually falling behind and becoming increasingly outdated. By the turn of the 21st Century, the 1968 Building Code had become an antiquated, complicated tangle of provisions.

In 2003, the Department of Buildings began a multi-year effort to replace the 1968 Building Code, culminating with Mayor Michael R. Bloomberg's signing of Local Law 33 of 2007. The result was the 2008 New York City Construction Codes, which replaced the 1968 Building Code with a new set of codes that increases public safety, incorporates the latest in engineering and technology, and contains progressive ideas on sustainable development. Most importantly, the new Construction Codes must be thoroughly reviewed and updated every three years, ensuring that New York City's construction regulations never again become outdated.

While the 2008 New York City Constructions Codes will apply to all new buildings beginning July 1, 2009, the 1968 Building Code, and its predecessor from 1938, will continue to remain relevant for years to come. First, certain new buildings filed prior to July 1, 2009 will continue to be subject to the 1968 code. Additionally, provisions of the 1968 code will apply to most alterations to existing buildings. Lastly, buildings constructed in accordance with the 1968 code generally must maintain compliance with its provisions.

The flowchart that follows the editor's note illustrates the circumstances under which the 1968 code remains applicable for alteration projects.

Robert D. LiMandri
Commissioner

EDITOR'S NOTE:

For further information, readers may wish to refer to the published series of the Department of Buildings' Directives and Memorandums, which are available at CityStore (NYC.gov/citystore) or visit the Department of Buildings website at NYC.gov/buildings for the latest policy and procedure notices.

The legislature enacted, effective September 1, 1986, Chapter 839 of the state laws of 1986, which made certain technical corrections and changes to the recodification.

Within the Reference Standards Appendix of this volume are references to specific sections in the Building Code. Standards enacted prior to the recodification of the Building Code refer to the code using the old section numbers. Editorial notes pointing out discrepancies between the former code and the recodified version not specifically indicated as changes, or references to laws that have amended the code since recodification, are indicated with asterisks and corresponding footnotes in bold italics at the following the section. Obvious errors (such as misspellings) are corrected and noted within the text with a [*sic*] following the particular word.

Page Setup:
Where text is interrupted by a table, left column above the table will continue unto the right column above the table. Text below the table will follow the same pattern.

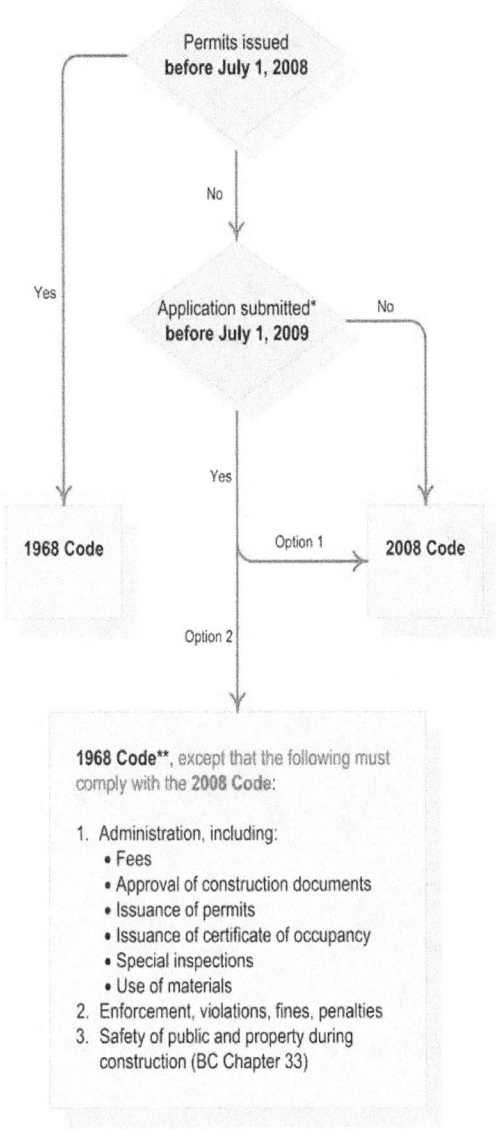

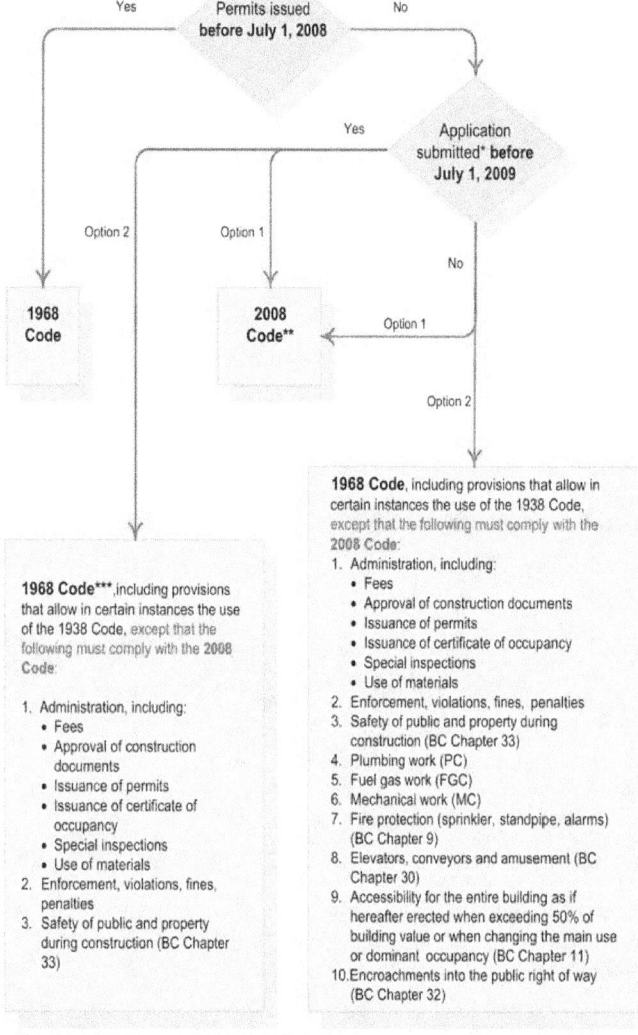

Which code applies?
Alterations to Existing Buildings

* Submission of application for construction document approval
** The 2008 Code cannot be elected where the 2008 Code provisions as applied to the particular building would result in a reduction in fire safety or structural safety. As an alternative, the entire building may be made to comply with 2008 Code
*** In addition, this option remains available only if:
    (1) the application is not abandoned;
    (2) the work is commenced within 12 months of issuance of a permit, and
    (3) the work is diligently carried out to completion

revision: July 1, 2008   **IIb**

SUBCHAPTER 3
BUILDING CONSTRUCTION

TABLE OF CONTENTS

Art.
[Sec.]* or Sec.**

Art. 1 General Provisions
[10.0] 205 Matters Covered
[11.0] 206 All New work to Conform

Art. 2 Permits
[20.0] 207 Requirement of Permit
[21.0] 208 Approval of Plans
[22.0] 209 Signature to Permit

Art. 3 Fees
[30.0] 210 Requirement of Fee
[31.0] 211 Fee for Approval of Plans and work Permits
[32.0] 212 Computation of Fees for work Permits
— 212.1 Civil penalty for work without a permit
[33.0] 213 Fees for Equipment Use Permits
[34.0] 214 Special Fees
[34.1] 214.1 Special Fees; Asbestos
[35.0] 215 Fees for the Testing, Approval, Inspection and Use of Power Operated Cranes, Derricks and Cableways

Art. 4 Inspections
[40.0] 216 Right of Entry and Inspection
[41.0] 217 Inspections of Building work
[42.0] 218 Inspection of Completed Buildings
[43.0] 219 Inspection of Construction Machinery and Equipment, Etc.
[44.0] 220 Inspection of Signs
[45.0] 221 Inspection Reports

Art. 5 Certificates of Occupancy
[50.0] 222 Requirement of Certificate of Occupancy
[51.0] 223 Occupancy of Existing Buildings
[52.0] 224 Issuance and Filing of Certificate of Occupancy

Art. 6 Projections Beyond Street Line
[60.0] 225 General Restrictions on Projections Beyond Street Line
[61.0] 226 Existing Projections Beyond Street Line
[62.0] 227 Rules Governing Projections Beyond Street Line

Art. 7 Safety in Building Operations
[70.0] 228 General Safety Requirements
[71.0] 229 Safety Requirements During Excavation Operations
[72.0] 230 Protection of Roofs, Skylights, Etc.
[73.0] 231 Regulation of Lots
[74.0] 232 Retaining Walls
[75.0] 233 Maintenance and Repair of Protection Fences and Retaining walls
[76.0] 234 Report and Inspection of Unsafe Buildings and Property

Art. 8 Unsafe Buildings and Property
[80.0] 235 Removal or Repair of Structures
[80.5] 236 Record and Notice of Unsafe Structures or Premises
[81.0] 237 Voluntary Abatement of Unsafe or Dangerous Conditions
[81.5] 238 Survey
[82.0] 239 Judicial Review of Survey
[82.5] 240 Repair or Removal Under Precept
[83.0] 241 Provision for Expense of Executing Precept
[83.5] 242 Return of Precept; Reimbursement of City
[84.0] 243 Fallen Structures and Structures Imminently Dangerous

Art. 9 Violations and Punishments
[84.5] 244 Notices of Requirements or of Violations
[85.0] 245 Emergency Measures
[85.5] 246 Judicial Remedies
[86.0] 247 Judicial Orders
[86.5] 248 Punishments
[87.0] 249 Violations of Peremptory Orders
[87.5] 250 Appeal

Art. 10 Miscellaneous Provisions
[91.0] 251 Illegal Practices in the Sale or Use of Lumber for Construction Purposes Prohibited
[92.0] 252 Sidewalk Sheds, Fences, Railings, Etc.

*"C26" omitted from section numbers in this column.
**"26" omitted from section numbers in this column.

ARTICLE 1
GENERAL PROVISIONS

§[C26-10.0] 26-205 Matters covered. All matters affecting or relating to the construction, alteration, repair, demolition, removal, maintenance, occupancy and use of new and existing buildings in the city, including the erection, installation, alteration, repair, maintenance, use and operation of signs and service equipment used in or in connection therewith, are presumptively provided for in this subchapter and in the building code of the city. This subchapter does not presumptively provide for matters that are contained in the charter, the labor law, the multiple dwelling law, subchapters one and two of this chapter and chapter two of title twenty-seven, the zoning resolution, or the general city law; nor does this subchapter apply to structures on waterfront property used in conjunction with and in the furtherance of waterfront commerce and/or

Title 26 / Subchapter 3

~~navigation, or to bridges, tunnels or subways, or to structures appurtenant thereto.~~

~~**§[C26-11.0] 26-206 All new work to conform.-**~~
~~All building work performed in the city on and after December sixth, nineteen hundred sixty-eight, shall conform to the provisions of this subchapter except that any work for which an application for a permit was filed prior to December sixth, nineteen hundred sixty-eight, and any work for which an application for a permit is filed within twelve months after the effective date of this subchapter, may be performed, at the option of the owner, in its entirety either in accordance with and subject to the requirements of this subchapter or in accordance with and subject to the requirements of the building laws and regulations previously in force in the city, provided that such work is commenced within twelve months after the date of issuance of a permit therefor and is continuously carried on to completion. This section shall not apply to the requirements of article ten of subchapter nineteen of title twenty-seven of the code which shall become effective December twenty-ninth, nineteen hundred sixty-nine.~~

~~**ARTICLE 2 PERMITS**~~

~~**§[C26-20.0] 26-207 Requirement of permit.-**~~
~~It shall be unlawful, on and after December sixth, nineteen hundred sixty-eight, to construct, alter, repair, demolish, or remove any building in the city, or to erect, install, alter, repair, or use or operate any signs or service equipment in or in connection therewith, unless and until a written permit therefor shall have been issued by the commissioner in accordance with the requirements of this subchapter and the requirements of the building code, subject to such exceptions and exemptions as may be therein provided.~~

~~**§[C26-21.0] 26-208 Approval of plans.-**~~
~~Whenever plans are required to be filed in connection with an application for a permit, as provided in the building code, all such plans shall be approved by the commissioner prior to the commencement of any work thereunder. All plans and all applications for approval thereof, shall comply with the requirements of the building code, subject to such exceptions and exemptions as may be therein provided; and all elevations on plans shall be referred to the United States coast and geodetic survey mean sea level datum of nineteen hundred twenty-nine, which is hereby established as the city datum.~~

~~**§[C26-22.0] 26-209 Signature to permit.-** Every permit issued by the commissioner shall have his or her signature affixed thereto; but the commissioner may authorize any subordinate to affix such signature.~~

~~**ARTICLE 3 FEES**~~

~~**§[C26-30.0] 26-210 Requirement of fee.-** No work permit or equipment use permit, when required by the provisions of the building code, shall be issued, and no plans or other statement describing building work, when required by the provisions of the building code, shall be approved, unless and until the required fee or fees therefor shall have been paid to the department in accordance with the provisions of this article, except that no fees shall be payable for work permits, equipment use permits or places of assembly permits if the owner of the building or property affected is a corporation or association organized and operated exclusively for religious, charitable or educational purposes, or for one or more such purposes, no part of the earnings of which enures to the benefits of any private shareholder or individual, and provided that the property affected is to be used exclusively by such corporation or association for one or more of such purposes.~~

~~*§[C26-31.0] 26-211 Fee for approval of plans and work permits.-~~
~~The fees required to be paid under this section, and under section 26-212 of this article, are for the filing and processing of applications for the approval of plans or other statement describing building work, the filing and processing of permit applications, the issuance or renewal of work permits, the inspection of building work, and the issuance of certificates of occupancy. Fifty percent of the total fee for the work permit, but not less than one hundred dollars, or the total fee for the work permit where such fee is less than one hundred dollars, shall be paid by or on behalf of the owner or lessee of the building premises or property affected and shall accompany the first application for the approval of plans or other statement describing the building work when submitted prior to submission of the permit application; and the whole or remainder of the total fee shall be paid before the work permit may be issued. A fee of one hundred dollars shall be paid with an application for renewal of a work permit. Foundation work, plumbing work, sign and service equipment work are included in the term "building" whenever plans for such work are required to be filed with construction or alteration plans; otherwise, separate fees shall be applied and collected for such work in accordance with the provisions of this section and section 26-212 of this article.~~
~~*Local Law 38-1990.~~

~~*§[C26-32.0] 26-212 Computation of fees for work permits.-~~
~~Fees for work permits shall be computed as hereinafter provided:~~
~~***1. New buildings.- The fees for permits to construct new buildings and open air stadia shall be computed as follows:~~
~~(a) Except as otherwise provided in paragraph (b),~~

revision: July 1, 2008   Strikethrough indicates repeal of text as per Local Law 33-2007   36
(See Title 28 of Administrative Code for new provisions)

Steven L. Carroll, P.E.
Consulting Engineer

Telephone 718-356-6245

226 Woodrow Road
Staten Island, New York 10312

February 6, 1985

Mr. Robert Diamond
Brooklyn Historical Railway Association
599 East 7th Street
Brooklyn, New York 11218

Dear Robert,

    Enclosed is a summary of my cost estimate with the appropriate backup calculations and references for all items of construction concerning the Atlantic Avenue Tunnel Project.

    I can be reached at the above address and telephone number if any questions arise or if my professional services are again needed. Good luck on this exciting project.

                                      Sincerely yours,

                                      Steven L. Carroll, P.E.

SLC:flc
Enc.

cc: S. Scialabba

Mr. Robert Diamond   -2-   February 6, 1985

Summary of Estimate

Mobilization	$ 15,000
Excavation	$ 277,531
Concrete	$ 106,883
Tunnel Work	$ 116,210
Restorations	$ 17,328
Removals	$ 14,050
Misc. Metal Work	$ 38,484
Moisture Protection	$ 6,667
Mechanical	$ 96,791
Electrical	$ 48,000
Utility Protection	$ 128,000
Communications	$ 4,000
Miscellaneous	$ 8,000
Borings and Test Pits	$ 3,446
Bond @ 1%	$ 8,804
Contingency @ 20%	$ 176,078
Adm. and Engineering Fees	$ 125,000
Total Cost of Project =	$1,190,272
SAY	$1.2 Million

Note that the entrance cost, which is included above, but consists of a variety of items, is estimated as follows:

    Court St. Entrance, $389,000

    Hicks St. Entrance, $352,000

Mr. Robert Diamond   -2-   February 6, 1985

<u>Summary of Estimate</u>

Mobilization	$ 15,000
Excavation	277,531
Concrete	106,883
Tunnel Work	116,210
Restorations	17,328
Removals	14,050
Misc. Metal Work	38,484
Moisture Protection	6,667
Mechanical	96,791
Electrical	48,000
Utility Protection	128,000
Communications	4,000
Miscellaneous	8,000
Borings and Test Pits	3,446
Bond @ 1%	8,804
Contingency @ 20%	176,078
Adm. and Engineering Fees	125,000

           Total Cost of Project = $1,190,272
                           SAY   <u>$1.2 Million</u>

Note that the entrance cost, which is included above, but consists of a variety of items, is estimated as follows:

     Court St. Entrance, $389,000

     Hicks St. Entrance, $352,000

CITY OF NEW YORK
DEPARTMENT OF ENVIRONMENTAL PROTECTION
BUREAU OF WATER SUPPLY

1250 BROADWAY, NEW YORK, N.Y. 10001   (212) 971-6796

JOSEPH T. McGOUGH, JR.
Commissioner

JOSEPH P. CONWAY, P.E., Director
Assistant Commissioner

September 13, 1984

Mr. Morris Tarshis, Director
Bureau of Franchises
Board of Estimate
Municipal Building, Room 1307
New York, New York 10007

      Re: Proposed Entrances to Railroad Tunnel
          on Atlantic Avenue, Borough of Brooklyn.

Dear Sir:

This is in response to your letter dated June 22, 1984 submitting a petition on the above subject. The Bureau was to review the petition to see if there were any objections to maintain and use an abandoned railroad tunnel, including the construction of entrances and facilities under, along and across Atlantic Avenue from east of Court Street to Hicks Street, Borough of Brooklyn.

The Bureau does have objections to these new entrances as shown on the plans. As proposed the petitioner would be required to replace sections of the sixteen (16) inch low pressure main on Court Street; the eight (8) inch low pressure main, the twelve (12) inch high pressure fire main, the twenty (20) inch low pressure main and finally the forty-eight (48) inch trunk main all crossing the proposed entrance at Atlantic Avenue. The cost of these replacements would be approximately one-hundred thousand ($100,000) dollars.

Therefore at Court Street it is suggested that the proposed entrance be placed along the south sidewalk of Atlantic Avenue east of Court Street. This would reduce the water mains to be replaced to sections of a twenty (20) and twelve (12) inch low pressure mains which approximately would cost sixteen thousand ($16,000) dollars.

Mr. Morris Tarshis 9/13/84 Page 2

At Hicks Street the new entrance would require the replacement of two (2) twenty (20) inch high pressure fire mains and one (1) twelve (12) inch low pressure main. The replacement costs of these would be eighteen thousand ($18,000) dollars. The southwest corner site for the proposed entrance does minimize the water main replacement.

As far as the proposed use of the manhole at Court Street this Bureau must object since there is a sixteen (16) inch high pressure water main exposed which would be in danger of being damaged. Therefore this manhole shall have to be removed and the area below backfilled to protect this main.

If the proposal is to be done this Bureau would require a bond to be posted for the replacement water main work. If the proposal is done as the petitioner decribes in his drawing the bond would be one-hundred eighteen thousand dollars ($100,000 for Court Street plus $18,000 for Hicks Street). If the petitioner changes the Court Street location as suggested to the south side of Atlantic Avenue the bond would be thirty-four thousand dollars (Court Street $16,000 plus Hicks Street $18,000).

All water main work shall be done under the inspection of Bureau inspectors after plans have been approved by the Construction Division of the Bureau.

Very truly yours,

Martin E. Engelhardt

MARTIN E. ENGELHARDT, P.E.
Chief, Planning & Programs
Bureau of Water Supply

GDeF/lb

bcc: Engelhardt, Dorf, Kushner, Kass, Brooklyn Borough Office
w/original submission

Schrader

FIRE DEPARTMENT
250 LIVINGSTON STREET   BROOKLYN, N.Y. 11201-5884

BUREAU OF FIRE PREVENTION

BOARD OF ESTIMATE
BUREAU OF FRANCHISES
RECEIVED
109193 SP 26 84

NUMBER ASSIGNED TO:
CHARNG-SHYONG WU

September 25, 1984

Mr. Morris Tarshis
Director Bureau of Franchises
Room 1307
Municipal Building
One Centre Street
New York, N.Y. 10007

Subject: Bureau of Franchise No. 108062

Dear Mr. Tarshis:

The Fire Department has reviewed the tentative proposal made by the Brooklyn Historic Railroad Association relative to the use of the abandoned railroad tunnel under Atlantic Avenue running from a point West of Boerum Place to a point East of Columbia Street, borough of Brooklyn.

This department will not oppose the project, provided that the safety of the public is paramount. We have discussed with Mr. Robert Diamond, President of the Brooklyn Historic Railroad Association, the requirements essential for public safety. These requirements include the submission of a Fire Protection Plan through the Department of Building from which the Fire Department will receive a copy for review and approval. We have outlined to Mr. Diamond the necessity for automatic sprinklers, a local alarm system, emergency lighting, standpipe, smoke and gas detectors, and forced ventilation system with proper controls and safeguards under fire conditions. We have also indicated the need for additional exit facilities suitable for the use of the public and in accordance with law.

Mr. Diamond has assured this department that he will meet the above requirements and, if possible, exceed them. Mr. Diamond will have a registered architect submit the required building and Fire Protection Plans through the proper channels outlined above. When suitable plans have been submitted and any recommendations for change have been complied with in the interest of public safety, the Fire Department will approve this project.

Very truly yours,

Robert E. Manson
Deputy Assistant Chief
Technology Management
Bureau of Fire Prevention

REM:MJB:mr

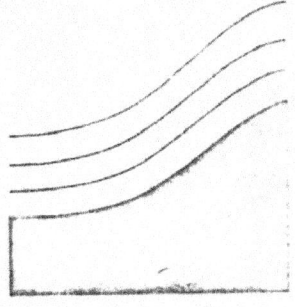

CITY OF NEW YORK
DEPARTMENT OF ENVIRONMENTAL PROTECTION
BUREAU OF SEWERS
40 WORTH STREET, NEW YORK, N.Y. 10013   (212) 566-2104/5

JOSEPH T. McGOUGH, JR.
Commissioner

AUG 23 1984

JOHN L. DiMARTINO, P.E., Director
Assistant Commissioner

Morris Tarshis, Director
Bureau of Franchises
Board of Estimate
Room 1307, Municipal Building
New York, NY 10007

Re: Petition by: The Brooklyn
Historic Railway Association,
Brooklyn, K235

Dear Mr. Tarshis:

This is in reply to your letter dated 22 June 1984 which referred to a petition by The Brooklyn Historic Railway Association requesting consent to maintain, and use an abandoned railroad tunnel, including the construction of an entrance and facilities to accommodate the public, under, along and across Atlantic Avenue from east of Court Street to Hicks Street, Borough of Brooklyn.

Please be advised that the Bureau of Sewers has no objection to the petitioner's request. This approval is predicated upon a similar favorable response from the Bureau of Water Supply which is required to comment separately.

Approval by the Department of Environmental Protection is valid only when approvals have been conveyed to your office by both the Bureaus of Water Supply and Sewers.

Very truly yours,

FRANK OLIVERI, P.E.
Acting Director
Bureau of Sewers

ROBERT M. LITKE
COMMISSIONER

CITY OF NEW YORK
DEPARTMENT OF GENERAL SERVICES
DIVISION OF PUBLIC STRUCTURES
MUNICIPAL BUILDING
16TH FLOOR
NEW YORK, N.Y. 10007

GREGORY JOHNSON
DEPUTY COMMISSIONER

September 12, 1984

Re: BROOKLYN HISTORIC RAILWAY ASSOCIATION
Petition dated June 18, 1984 to the
BOARD OF ESTIMATE for consent to
maintain and use an abandoned railroad
tunnel, including the construction of an
entrance and facilities to accommodate
the public under, along and across
Atlantic Avenue from east of Court Street
to Hicks Street
Bureau of Franchises #108062
Borough of Brooklyn

Mr. Morris Tarshis
Director
Bureau of Franchises
1307 Municipal Building
1 Centre Street
New York, N.Y. 10007

Dear Mr. Tarshis:

This is in reply to your June 22, 1984 letter regarding the above matter.

Please be advised that we have no objections to the above consent.

Very truly yours,

Martin Burrell, P.E.
Director
Bureau of Electrical Control

HT/EC:341/ip

BOARD OF ESTIMATE
BUREAU OF FRANCHISES
RECEIVED
108076 SP 14 84

NUMBER
ASSIGNED TO: CHARNG-SHYONG WU

CITY OF NEW YORK
DEPARTMENT OF TRANSPORTATION
OFFICE OF THE COMMISSIONER
40 WORTH STREET • NEW YORK, N.Y. 10013

ANTHONY R. AMERUSO, P.E.
COMMISSIONER

OCT 1984

BK-DOM-3S-C-600

CHARNG-SHYONG WU

Honorable Morris Tarshis
Director of Franchises
Board of Estimate
Municipal Building
New York, New York 10007

Dear Mr Tarshis:

Regarding the petition to the Board of Estimate from The Brooklyn Historic Railway Association dated June 13, 1984 requesting consent to construct, maintain and use an abandoned railroad tunnel, including the construction of an entrance and facilities to accommadate the public, under, along and across Atlantic Avenue from East of Court Street to Hicks Street, in the Borough of Brooklyn, please be advised that the Department of Transportation has no objection to the petition provided the following conditions are observed:

1) Permits are secured from the Brooklyn Borough Office before starting work.

2) The applicant will restore any existing curb, sidewalk or roadway damaged during construction in accordance with Bureau of Highway Operations Standard Specifications.

3) Within thirty (30) days of completion of construction the petitioner shall submit one set of certified revised "as built drawings" and microfiche card of drawings and related correspondence. Certification to be made by a licensed P.E.

4) The petitioner shall notify utility companies and agencies having existing underground facilities in the proposed construction area for their review and approval.

2

5) The applicant shall comply with applicable sections of Industrial Code Rule 53 of the State of New York (construction, Excavation and Demolition Operations at or near underground facilities).

6) Inspection Reports must be filed with the Bureau of Highways-Operations Mangement at five year intervals certifying the following:

   A) The structural members were inspected by the engineer within the last six (6) months.

   B) The load carrying capacity is sufficient to support the anticipated loading.

   C) The non-load carrying members have been inspected and are secure.

   D) In addition, a microfiche card of above mentioned "as built drawings" and of related correspomdence shall be submitted by the petitioner.

Very truly yours,

Anthony R. Ameruso, P.E.
Commissioner

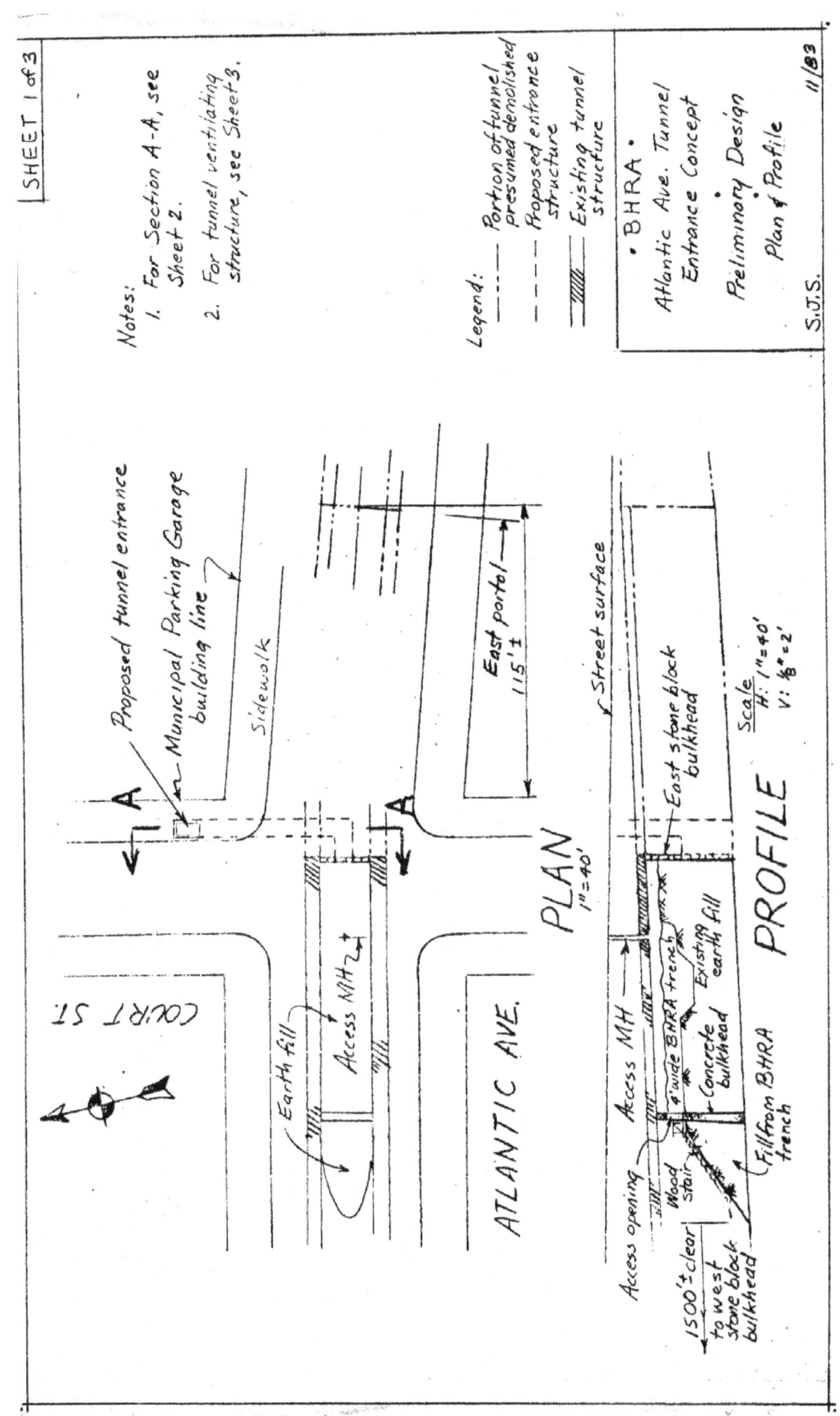

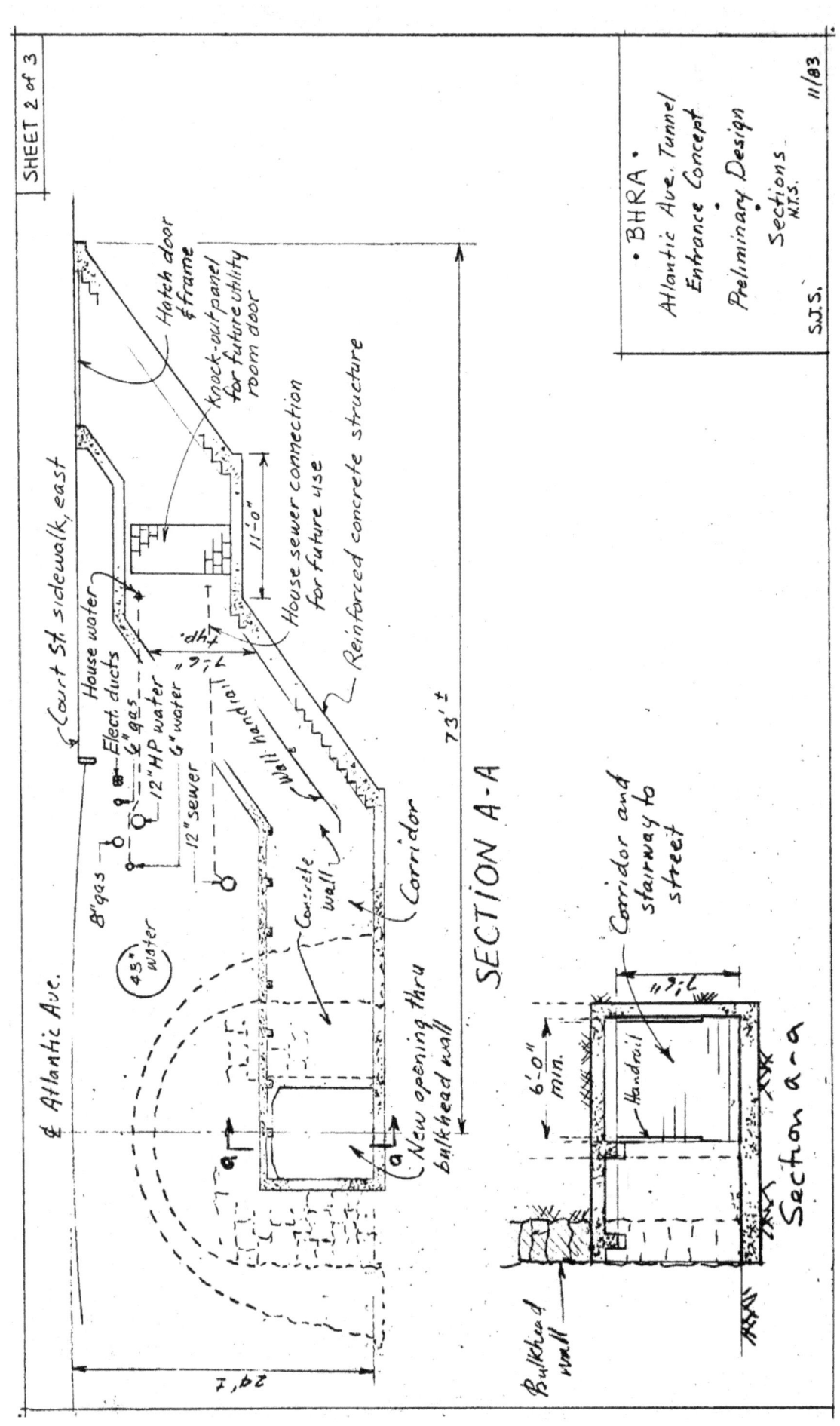

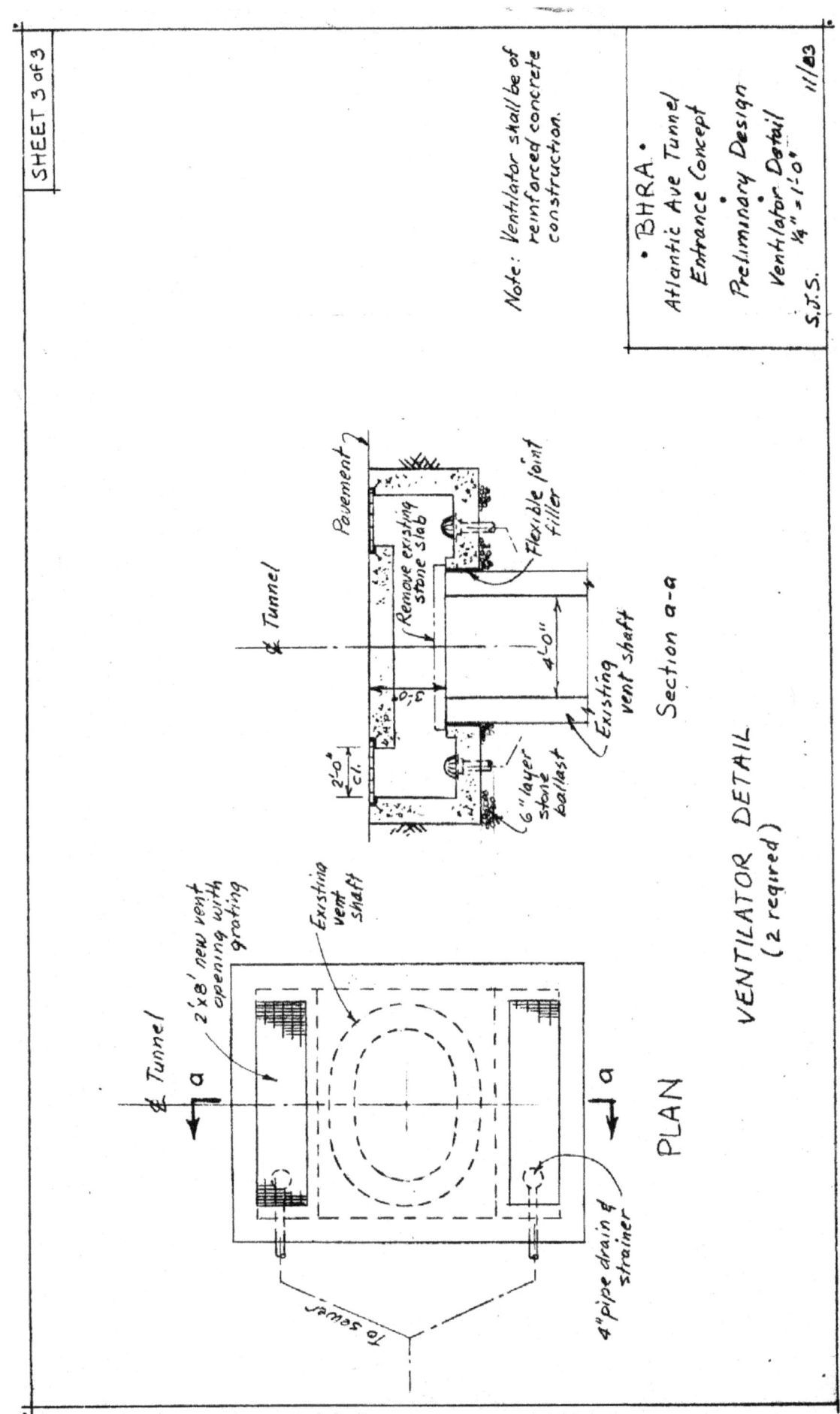

Item		Description		Unit	Quantity	Price	Amount
1.		Mobilization		L.S.	—	—	$15,000
2.		Sheeting		Sq. ft.	7,260	28.51	$206,983
3.		Decking	Timber	B.M.	22,303	1.59	$35,462
			Steel	lbs	9,000	.96	$8,640
4.	Excavation	Earth Excavation		cu.yd	1,476	7.55	$11,147
5.		Excav. of Exist Tunnel Fill		cu.yd	195	49	$9,575
6.		Stone Ballast		cu.yd	56	27.80	$1,577
7.		Backfill		cu.yd	650	6.38	$4,147
8.	Concrete	Concrete for Entrances		cu.yd	255	381	$97,155
9.		Concrete for Vents		cu.yd	28	276	$7,728
10.		Steel Reinforcing		lbs	28,454	.65	Included
11.	Tunnel Work	Sleeves in Concrete		each	100	20	$2,000
12.		Tunnel Flooring		Sq. ft.	10,000	2.00	$20,000
13.		Handrail		L.F.	1,000	16.21	$16,210
14.		Railroad Track		L.F.	1,000	80	$80,000
15.	Restorations	Saving Excav. Tunn. Fill		Stone	Unknown	5.00	unknown
16.		Record. Stone Block Walls		Sq. ft.	1,268	2.70	$3,424
17.		Entrance Sculpture		L.S.	2	5,000	$10,000
18.		Restoration of Sidewalk		Sq. ft.	840	3.49	$2,932
19.		Restoration of Pavement		Sq. Yd.	134	7.25	$972
20.	Removals	Concrete Excavation		cu.yd	36	183	$6,588
21.		Pavement Excavation		sq.yd	215	4.90	$1,054
22.		Remove Stone Block Wall		cu.ft.	600	3.07	$1,842
23.		Remove Concrete Bulkhd		cu.ft.	1,512	3.02	$4,566
24.	Misc. Metal Work	Entrance Handrails		L.F.	160	34.99	$5,598
25.		Non slip Stair Treads		L.F.	608	12.97	$7,886
26.		Entrance Hatch Doors		each	2	7,000	$14,000
27.		Vent Stacks @ Entrance		L.F.	56	89.30	$5,000
28.		Steel Ladders @ Vents		lbs	6000	1.00	$6,000
29.	M.P.	3 ply Waterproofing		Sq. ft.	1,692	1.69	$2,860
30.		4 ply Waterproofing		Sq. ft.	1,692	2.25	$3,807
31.	Plumbing	Fire Line Standpipes		L.F.	1,663	34	$56,542
32.		Exhaust Fans		each	3	4,083	$12,249
33.		Plumbing		L.S.	—	—	$28,000
34.	Electrical	Electrical a) lighting		L.S.	—	—	$48,000
		b) Emergy light					
		(included above) c) clock outlets					
		d) Emerg. Alarm					

Item	Description	Unit	Quantity	Price	Amount
35.	Replacement of Mains	L.S.	——	——	$118,00
36.	Maint. and Supp (Utilities	L.S.	——	——	$10,000
37.	Public Address	L.S.	——	——	$3,000
38.	Telephone	L.S.	——	——	$1,000
39.	Crane to Lower Trolley	L.S.	——	——	$1,000
40.	Refr, Souvenir, + Exhibit Stds	L.S.	——	——	$7,000
41.	Borings (4) + Test pits (8)	L.S.	——	——	$3,446
42.	Bond @ 1%	L.S.	——	——	$8,804
43.	Contingency (20%)	L.S.	——	——	$176,07
44.	Adm. and Engineering Fees	L.S.	——	——	$125,00
			TOTAL	=	$1,190,2
45	Exterminator	?			?

STATE OF NEW YORK
STEVEN L CARROLL
052425
LICENSED PROFESSIONAL ENGINEER

Costs

1982 Means   Steven L. Carroll
       12/5/83

Concrete:

     Bms — 2 cuyds
     Roof — 27 cuyds

                                       M fact escal

82   29 cuyds × $345/cuyd × 1.15 × 1.2   = $13,807
:3-275

82   Mat — 37 cuyds × $190/cuyd × 1.15 × 1.2 = $9,701
3-400

12   Walls — 46 cuyds × $300/cuyd × 1.15 × 1.2 = $19,044
:3-427

         Vent — 28 cuyds × $200/cuyd × 1.15 × 1.2 = $7,728

                                             Total = $50,280

Sheeting:

33   3080 sq.ft. × $22/sq ft × 1.08 × 1.2 = $87,817
3-40-080

Excavation:

         (+ 54 cuyds)

29   609 cuyds × $1.98/cuyd × 1.08 × 1.2 = $1,563
:3-025   heavy traffic   +$139

     hauling —   (1.3×)   +$269

32   609 cuyds × ($2.96/cuyd × 1.08 × 1.2 = $3,037
3-050   (+ 54 cuyds) $4,600 + $408
+140   = $5,008

Backfilling:

27   264 cuyds × $4.92/cuyd × 1.08 × 1.2 = $1,683
3-200

Stone Ballast:

P.31
2.3-110
 & 320

$26 \text{ cuyds} \times \$14.30/\text{cuyd} \times 1.08 \times 1.2 = \$723$

(1.5×) — compaction

Pavement Removal:

P.26
2.1-170
~P.25

$95 \text{ sqyds} \times \$1.26/\text{sqyd} \times 1.08 \times 1.2 = \$465$

(3×) — congested site + small amt.

Concrete Removal:
(Street)

P.25
2.1-030
 & -039

$16 \text{ cuyds} \times 3 \times \$47/\text{cuyd} \times 1.08 \times 1.2 = \$2,924$

Roadway Decking:
(12"×12" Timbers)

P.153
6.1-140

$810 \text{ cu.ft.} \times \dfrac{\text{LBF}}{.0825} \times \$1.18/\text{BF} \times 1.12 \times 1.2 = \$15,571$

(12 WF 25z)

P.131
6.1-151

$4,250 \text{ lbs}/2000 \text{ lbs} \times \$1,475/\text{ton} \times 1.08 \times 1.2 = \underline{\$4,062}$

$\$19,633$

Removal of Concrete Bulkhead:

P.25
2.1-010

$(756 \text{ cuft})/27 \times \$63/\text{cuyd} \times 1.08 \times 1.2 = \$2,286$

Removal of Stone Wall:

P.26
2.1-140

$300 \text{ cu ft} \times \$2.37/\text{cu.ft.} \times 1.08 \times 1.2 = \$921$

Tunnel Flooring: (OMIT Calculations Below)
(4"×8") — See Pages 10 + 11 —

P.158
2-020

$760 \text{ L.F.} \times \$4.07/\text{ft} \times 1.12 \times 1.2 = \$4,157$

(plywood) 5/8"

P.156-010

$1200 \text{ sq.ft.} \times \$.71/\text{sq ft} \times 1.12 \times 1.2 = \$1,145$

$\}\ \$5,302$

Reinforcement:

P.78
3.2-060
Mat — $4,534\, lbs / 2000\, lbs \times \$915/ton \times 1.18 \times 1.2 = \$2,937$

P.78
3.2-040
Roof — $2,748\, lbs / 2000\, lbs \times \$840/ton \times 1.18 \times 1.2 = \$1,634$

P.77
3.2-010
Beams — $2,975\, lbs / 2000\, lbs \times \$1,100/ton \times 1.18 \times 1.2 = \$2,317$

P.78
3.2-070
Walls — $3,142\, lbs / 2000\, lbs \times \$835/ton \times 1.18 \times 1.2 = \underline{\$1,858}$

$$\text{Total} = \$8,746$$

Restoration of Sidewalk:

(6")
P.49
2.6-040
Concrete — $420\, sqft \times \$2.37/sqft \times 1.08 \times 1.2 = \$1,290$

P.49
2.6-045
(4")
Stone — $420\, sqft \times \$.32/sqft \times 1.08 \times 1.2 = \underline{\$174}$

$$\$1,464$$

Restoration of Pavement:

P.47
2.6-010
$67\, sqyds \times \$5.60/sqyd \times 1.08 \times 1.2 = \$486$

Handrail:

P.138
5.4-240
$80\, ft \times \$27/ft \times 1.08 \times 1.2 = \$2,799$

Stair Treads:

(4" wide)
P.86
3.3-002
$38\, tr \times 8\frac{ft}{tr} \times \$9.40/ft \times 1.15 \times 1.2 = \$3,944$

Recondition Stone Block Wall:

P.322
18.1-055

634 sq ft. × $2.25/sq ft. × 1.0 × 1.2 = $1,712

Excavating Tunnel fill:

Hand excavate –

P.30
2.3-140
+150
P.8
1.5-020

Inside cover

195 cu yds × $30/cu yd × 1.08 × 1.2 = $7,582

Backhoe Rental –

1 Day/36 cu yds × 195 cu yds × $5,650/mo × 1/30 days

out put of hand Excavators above (use 6 Laborers) ∴ daily output = 36 cu yds

= $1,020

Labor (oper Eng + oiler) –

1 day/36 cu yds × 195 cu yds × $25.65/hr × 7 hrs/1 day

= $973

Total = $9,575

---

3 × 4 ply water proofing:

P.172
7.1-070

3 ply
736 sq ft. × $1.42/sq ft. × .99 × 1.2 = $1,242

4 ply
736 sq ft × $1.89/sq ft. × .99 × 1.2 = $1,653

$2,895

Borings:

P.24
2.1-080

2 × 50' × $9.45/ft. × 1.08 × 1.2 = $1,225

Test Pits:

P.26
2.1-050
+010

Avg 4 × (3' × 3' × 6')/27 × $48/cu yd × 1.08 × 1.2 = $498

Electrical Work:

$244,381 × $200,000 (Elect. work on similar proj.) / $3,000,000 (Gen Const. Cost on similar proj.) ≈ $16,000
(Gen Const. Cost) This project

Plumbing Work:

$244,381 × $121,000 (") / $3,000,000 (") ≈ $9,000

<u>USE $25,000</u>

Public Address:

P. 355
16.8-030 Microphone   - 1 × $65   → $65
 " -040 Speakers   - 3 × $72   → $216
 " -100 Monitor Panel - 1 × $185 → $185
 " -140 Volume Control - 1 × $51 → $51
 " -160 Amplifier   - 1 × $925 → $925
 " -180 Cabinets   - 1 × $525 → $525

$1,967 × 1.06 × 1.2 = $2,502

<u>USE $3,000</u>

(For total see Summary Sheet)

1/10/84

Steven L. Carroll

<u>1982 Means</u>
Reference for prices are in the "costs" section of these calculations

<u>2nd Entrance</u>

Concrete — $42,552

Sheeting — $87,817

Excavation — $4,600

Backfilling — $1,683

Stone Ballast — $123

Pavement removal — $553

Concrete removal — $2,924

Roadway Decking — $19,633

Removal of concrete bulkhead — $2,286

Removal of Stone wall — $921

Restoration of Sidewalk — $1,464

Restoration of pavement — $486

Handrail — $2,799

Stair treads — $3,944

3 & 4 ply waterproofing — $2,895

Borings — $1,225

Test pits — $498

Electrical = $48,000

Plumbing = $28,000

Additional ~~30~~ 22 ft. of corridor —

Concrete

Mat — $\dfrac{1' \times 22' \times 11'}{27} \times \$190/\text{cuyd} \times 1.15 \times 1.2 = \$2350$

Bms — $4\,Bms \times \dfrac{9'' \times 7.5' \times 8''}{12 \times 27 \times 12} \times \$345/\text{cuyd} \times 1.15 \times 1.$

$= \$26.$

Roof — $\dfrac{1' \times 10' \times 22'}{27} \times \$345/\text{cuyd} \times 1.15 \times 1.2 = \$3879$

Walls — $\dfrac{1' \times 8.25 \times 22'}{27} \times 2 \times \$300/\text{cuyd} \times 1.15 \times 1.2 = \$556$

Sheeting

32 ft. drive for 25 ft. excavation —

$25\,ft. \times 22\,ft. \times 2\,sides \times \$22/sqft. \times 1.08 \times 1.2 = \$31,363$

Excavation

$\dfrac{10\,ft. \times 25\,ft. \times 22\,ft.}{27} \times (\$1.98 + 1.3 \times \$2.96)/\text{cuyd} \times 1.2 \times 1.08 = \$1539$

Backfill

$$\frac{15\,ft. \times 10\,ft. \times 22\,ft.}{27} \times \$4.92/cuyd \times 1.08 \times 1.2 = \underline{\$779}$$

Stone Ballast:

$$\frac{6" \times 10' \times 22'}{12 \times 27} \times \$14.30/cuyd \times 1.5 \times 1.08 \times 1.2 = \underline{\$113}$$

Pavement Removal

$$\frac{10' \times 22'}{9} \times \$1.26/cuyd \times 3 \times 1.08 \times 1.2 = \underline{\$120}$$

Concrete Removal

$$\frac{6" \times 10' \times 22'}{12 \times 27} \times \$47/cuyd \times 1.08 \times 1.2 = \underline{\$248}$$

Roadway Decking

Timber 2×12
$$\frac{10' \times 22' \times 1\,ft.}{.0825\,cu.ft./B.F.} \times \$1.18/B.F. \times 1.12 \times 1.2 = \underline{\$4,229}$$

2WF25
$$\frac{2\,BMS \times 10' \times 25\,plf}{2000\,lbs} \times \$1,475/ton \times 1.08 \times 1.2 = \underline{\$478}$$

$$\underline{3 \text{ \& } 4 \text{ ply waterproofing}}$$

$$10 \text{ ft.} \times 22 \text{ ft.} \times \$1.42/\text{sq.ft.} \times .99 \times 1.2 = \underline{\$371}$$

(Total for 22 ft. of Additional Corridor)

$$= \underline{\$51,300}$$

(Total for Duplicate or 2nd Entrance) —

$$= \underline{\$224,003}$$

ToT (Entire 2nd Entrance) = $\underline{\underline{\$275,303}}$

Tunnel Flooring

<u>Delete</u> the item "Tunnel Flooring" in the previous calculations in the "Costs" section $-(\underline{\$5,302}) + 20\% \text{ (cont.)} = (\underline{\$6,362})$

<u>add the following</u>:

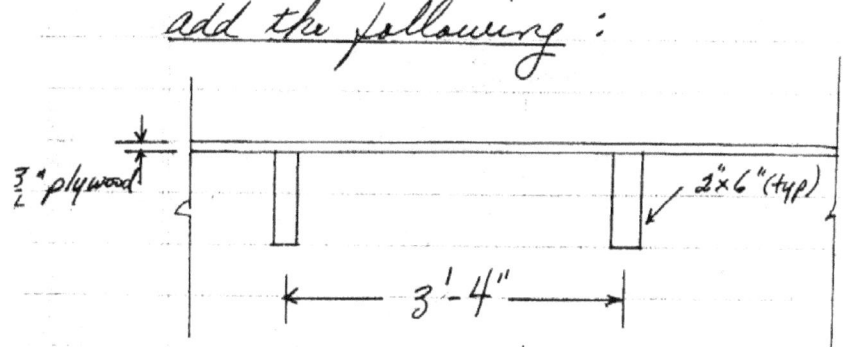

<u>Years</u> Platform 10 ft. wide × 1000 ft. long
1982

2"×6"'s – 3'-4" o.c. or 4 in 10 ft.

4 × 1,000 ft. = 4,000 L.F. or B.F.

<u>3/4" plywood</u>

10 ft. × 1,000 ft. = 10,000 sq. ft.

                          (NY fact) (escal) to 1984

P. 152 $.90/BFT × 4,000 BFT × 1.12 × 1.2 = $4,838
6.1-410

P. 155 $.82/sq.ft. × 10,000 sq.ft. × 1.12 × 1.2 = <u>$11,021</u>
6.1-030

                    <u>Total for Tunnel flooring</u> = $15,859
                    without cost of creosote
                         or fire retarder

11

Wood treatments:

(material only) —

Fire Retardant —

P.154 2"×6", $\$.150/B.F. \times 4,000\ B.F. \times 1.12 \times 1.2 = \$806$
6.1-040

P.154 3/4" Ply, $\$.18/sq.ft. \times 10,000\ sq.ft. \times 1.12 \times 1.2 = \underline{\$2,419}$
6.1-010

$$\text{Fire Ret. TOT} = \$3,225$$

Creosote —

P.154 2"×6", $\$.15/B.F. \times 4,000\ B.F. \times 1.12 \times 1.2 = \underline{\$806}$
6.1-030

$$\text{ToT. Treatment} = \underline{\$4,031}$$

$$\underline{\text{ToT for Treated Tunnel Floor}} = \$19,890$$

Metal Railing (mild steel, economy)

P.138 $\$12.55/ft. \times 1,000\ ft. \times 1.076 \times 1.2 = \underline{\$16,205}$
5.4-191

N.Y.C. Building Code
Standards for Natural Ventilation
(Article 12, Subarticle 1205.0)

Floor Area:

Tunnel — 21 ft. × 1,667 ft. = 35,007 sq. ft.
Entrances — (6 ft. × 76.75 ft. + 8 ft. × 7.5 ft.) × 2
        + 7.5 ft. × 22 ft. = 1,206 sq. ft.

    Total Floor Area = 36,213 sq. ft.

paragraph C-26-1205.5

Min Ventilation Area:

A_{min} = .05 × 36,213 sq. ft. = 1,811 sq. ft.

Para C26-1205.6 — ½ of above area is permissable if mechanical means of Ventilation is used:

∴ Min Ventilation Area = $\frac{1,811 \text{ sq. ft.}}{2}$
                                = 906 sq. ft.

Area Furnished:

Tunnel Vents = 3π ab (ellipse) = 3 Vents × π × 4 ft. × 6 ft.
                = 226 sq. ft.
2 Entrance doors = 2 × 12.5 ft. × 8 ft. = 200 sq.

Para C26-1205.6 — Min Vent opening = 3 sq. ft.

∴ Install 4 Vents – 2 ft. ⌀ each
— Ductile Iron —

$A = \dfrac{\pi D^2}{4} = \dfrac{\pi (2ft)^2}{4} = 3.14$ sq.ft. > 3 sq.ft.
                                                                                                             O.K.

$A_{(vents)} = 4 \times 3.14$ sq.ft. $= 12.6$ sq.ft.

Total Vent. Area Furnished $= 438.6$ sq.ft. > 906
                                                                                                            N.G.

∴ Use Mechanical means as supplemental Ventilation

Standards for Mechanical Ventilation

Article 12, Subarticle 1206.0

Vent. Index:
Para C26-1206.2   Max Occup. $= \dfrac{A_{min} \times 200}{V.I.}$

    Max Occup $= \dfrac{906 \text{ sq.ft.} \times 200}{301} = \underline{601 \text{ persons}}$

Para C26-1206.3, table 12-2

  Req'd Supply $= 2$ cfm/sq.ft.

    $Q_{in} = 2$ cfm/sq.ft. $\times 36,212$ sq.ft. $= 72,424$ cfm

Req'd Exhaust = 1.5 cfm/sq.ft. × 36,212 sq.ft.

$\qquad$ = 54,318 cfm

Q_{out} = 54,318 cfm req'd

<u>furnished</u>:

3 fans @ Q_{out} = 21,600 cfm each

Q_{out} furnished = 3 × 21,600 cfm = 64,800 cf.

$\qquad$ > 54,318 cf. req'd

O.K.

Exhaust fans

21,600 CFM, 40 sq in damper

P. 327
5.7-730
   3 fans × $3,250/fan × 1.047 × 1.2 = $12,250

Railroad Track

100 lb rail
6" × 8" × 8'-6" wooden ties
6" of stone ballast

P. 65
27
   $58.05/L.F. × 1000 ft. × 1.077 × 1.2 = $75,024

$$2^{nd} \text{ distribution, add} \longrightarrow \$5,000$$

$$\text{TOT R.R. Track} = \$80,000$$

Utility connections

water and electric hook up - $1000 each

$$\text{TOT} = \$2,000$$

Vent Stacks

P. 44
2.5-149
   4 Vents × 14 ft./Vent × $59/ft × 1.077 × 1.2 = $4270

plates, bolts, etc - say $5,000

Sprinkler Requirements
(Art. 3)

occupancy classification — F-3

as per subarticle 307.0, para c26-307.4

construction group — I-A or I-B or at least I-C

as per sub art — 314.0, paras c26-314.2, .3, .4

No Sprinkler Req'd as per

Article 4, table 4-1, Page 4-4

Fire Standpipe Requirements
(Article 17)

No. of Standpipes required — subarticle 1702.C
para c26-1702.4

125 ft. of hose (max.)
20 ft. of water stream (min)

Total radius = 145 ft.

distance in both directions from standpipe
= 290 ft. between standpipes (MIN)

2 entrances + tunnel = 1,751 ft.

No. of Std pipes Req'd = $\frac{1,751 \text{ ft.}}{290 \text{ ft./stpipe}}$ = 6

Size of Standpipe Req'd
(article 17)

Subarticle 1702.0, para C26-1702.7, and table 17-1

height < 150 ft. ∴ <u>Min ⌀ = 4"</u>

Reference Standard RS-17
  table RS 17-1-1

height < 115 ft.

∴ Use - 2½" hose & outlet valves, class A
  pipe - Schedule 40 mild steel
  fittings - steel (350 psi)
  Valves - check or gate valves (150 psi)

Design

6 class A hoses, racks, etc
1,645 ft. of schedule 40 steel pipe (4"⌀)
10 steel tees (4"⌀)
10 steel elbows (4"⌀)
18 ft. of branch pipe 4"⌀ sched 40
7 - 4" steel valves

Pour 1 ft. × 1 ft. cradle (conc.) 1,600 long

Fire Standpipe

P.293 $(1,645 + 18)\text{ ft.} \times \$14.15/\text{ft.} \times 1.047 \times 1.2 = \$29,565$
15.1-191

P.293 $10 \times \$50/\text{tee} \times 1.047 \times 1.2 = \$628$
15.1-480

P.293 $10 \times \$35/\text{elbow} \times 1.047 \times 1.2 = \$440$
15.1-480

P.299 $7 \times \$840/\text{valve} \times 1.047 \times 1.2 = \$7,388$
15.1-195

P.307 $6 \times \$180/\text{hose cabinet} \times 1.047 \times 1.2 = \$1,357$
15.4-410

P.308 $125 \text{ ft. of Hose} \times 2 \times 6 \times \$1.80/\text{ft.}$
15.4-238 $\times 1.047 \times 1.2 = \$3,392$

P.82 $1600 \text{ ft. (conc. cradle)} \times \dfrac{1' \times 1' \times \$170/\text{cu yd}}{27 \text{ ft.}}$
3.3-390
$\times 1.152 \times 1.2 = \$13,926$

Total = $\$56,696$

Quantities

Steven L. Carroll
12/5/83

Concrete: TOTAL — 126 cu yds
(Incl. vent shaft)

__Mat__
1st Stair —
1' x 21' x 11' = 231
1st Landing —
1' x 10' x 11' = 110
2nd Stair —
1' x 18.6' x 11' = 205
2nd Landing —
1' x 32' x 11' = 352
Entrance —
1' x 8' x 11' = 88
         986/27 ⟶ 37 cu yds

__Roof__
1st Stair —
1' x 6' x 10' = 60
1st Landing —
1' x 10' x 10' = 100
2nd Stair —
1' x 18.6' x 10' = 186
2nd Landing —
1' x 28' x 10' = 280
Entrance —
1' x 8' x 10' = 80
Drop —
2 x 10' x 1' = 20   726/27 ⟶ 27 cu yds
Beams — 8x18", 12" incl above, use 6" x 8"
19 bms x (6"x8")/144" x 8' = 51/27 ⟶ 2 cu yds

__Walls__
1st Stair —
(14½' x 12')/2 + 2' x 12' = 111
1st Landing —
10' x 8' x 1' = 80
2nd Stair —
15 x (8' + 8')/2 x 1' = 120
2nd Landing —
8' x 22' x 1' = 176
Entrance —   (seperate)   551 x 2/27 = 41 cu yds
8' x 8' x 1' = 64, 8 x 10 x 1 = 80, 8 x 8 x 1 = 64   + 5 cu yds

Reinforcement: **TOTAL – 13,400 lbs**

Mat –

            For
            (splices)   (#5)   Top & bott

Long – 89.6′ × 1.1 × 11 bars × (#5) 1.043 plf × 2 = 2,262 lbs
Short – 11′ × ″ × 90 bars × (#5) 1.043 ″ × 2 = 2,272 lbs
                                                                                   4,534 lbs

Roof –

           (splices)   (#4)   Top & bott

Long – 73′ × 1.1 × 10 bars × (#4) .668 plf × 2 = 1,073 lbs
Short – 10′ × ″ × 73 bars × (#5) 1.043 plf × 2 = 1,675 lbs
                                                                                2,748 lbs

Beams –

19 beams × 8′ × 4 bars × (#9) 3.4 plf = 2,067 lbs
10′/8/12 × 52″/12 × 19 beams × (#4) .668 plf = 908 lbs
                                                   × 1.1
                                                         2,975 lbs

Walls –

1st Stair –
Horiz – 16½′ × 12 bars × (#5) 1.043 plf × 1.1 = 110 lbs
Vert – 13½′ × 16 bars × (#5) 1.043 plf × 1.1 = 119 lbs

1st Landing –
Horiz – 10′ × 8 bars × (#5) 1.043 plf × 1.1 = 92 lbs
Vert – 9′ × 10 bars × (#5) 1.043 plf × 1.1 = 103 lbs

2nd Stair –
Horiz – 18.6′ × 8 bars × (#5) 1.043 plf × 1.1 = 171 lbs
Vert – 9′ × 19 bars × (#5) 1.043 plf × 1.1 = 196 lbs

2nd Landing –
Horiz – 32′ × 8 bars × (#5) 1.043 plf × 1.1 = 294 lbs
Vert – 9′ × 32 bars × (#5) 1.043 plf × 1.1 = 330 lbs

Entrance –
Horiz – 8′ × 8 bars × (#5) 1.043 plf × 1.1 = 73 lbs
Vert – 9′ × 8 bars × (#5) 1.043 plf × 1.1 = 83 lbs
                                                              1571 × 2 = 3,142 lbs

<u>Sheeting:</u>   <u>Total – 3080 Sq.ft.</u>

1st Stair –
(15½' × 14')/2 + 2'×14' = 137 sq.ft.
1st Landing –
8' × 14' = 112 sq.ft.
2nd Stair –
(14' + 26')/2 × 15' = 225 sq.ft.
2nd Landing –
22' × 26' = 572 sq.ft.
Entrance –
8' × 26' = 208 sq.ft.   1,254 sq ft × 2 = 2,508 sq.ft.
10' × 26' = 260 } separate
10' × 26' = 260 }   → + 572 sq.ft.
2 × 26' = 52

<u>Roadway Decking:</u>

$\frac{12" \times 12"}{144 \text{ ft}^2}$ × 10' × 81 timbers = 810 cu.ft.

≈ 170' × 12WF 25 plf sect = 4,250 lbs

<u>Concrete for Vent Shaft:</u>

52' × 4.3' × 1' = 224 cu ft
9' × 9' × 1' = 81 cu ft
2 × 3½' × 9' × 1' = 63 cu ft
4"/12 × (16' + 20') × 1' = <u>12 cu ft</u>
        380/27 = 14 cu yds × 2 vents = 28 cu y

<u>Excavating Existing Tunnel Fill:</u>   <u>Total = 195 cu yds</u>

Between Bulkheads –
½ (20' + 20') × 100' × 21' = 4200 cu ft
After 2nd Bulkhead –
(15' + 8')/2 × 20' + (8×21)/2 × 20/2 = <u>1070 cu ft</u>
                        $\frac{5270 \text{ cu ft}}{27}$ = 195 cu yd

3 & 4 ply Waterproofing:

$10' \times (6' + 10' + 18.6' + 28' + 8' + 3') = 736$ sq.ft.

Restoration of Sidewalk:

$12' \times 35' = 420$ sqft

Restoration of Pavement:

$50' \times 12' = 600$ sq.ft. $\times 1/9$ sq.ft. $= 67$ sq yds

Handrail:
(3 ft high)

2 sides $\times (18.6' + 21') \cong 80$ ft.

Non Slip Stair Treads:
(8' Long)

$31' / (10''/12) \cong 38$ treads

Removal of concrete Bulkhead:

$(3'+1')/2 \times 18' \times 21' = 756$ cu ft

Entrance opening in Stone wall:

$10' \times 10' \times 3' = 300$ cu ft.

Recondition Stone Block Walls:

$[9' \times 21' + (\pi \times 21^2)/8] \times 2$ walls $= 724$ sq ft $- 90$ sq ft (opening)
$= 634$ sq. ft.

Entrance sculpture:
76.5

$[((5' + 3\frac{1}{2}')/2) \times 9' \times 2 + \pi(20^2 - 8^2)/8] \times 2$ walls $= 237$

OMIT ABOVE Calculations
Tunnel Flooring: & Below

$31 \times (4'' \times 8'')/144 \times 18.7' + 2 \times (4'' \times 8'')/144 \times 60' =$ (USE 760 L.F.

Plywood 5/8" — $60' \times 20' = 1200$ sqft

Backfill: Total = 264 cu yds

1st Landing –
10' x 3' x 12' = 360 cu ft

2nd Stair –
(3' + 15')/2 x 12' = 108 cu ft

2nd Landing –
15' x 29' x 12' = 5220 cu ft

Entrance –
15' x 8 x 12' = 1440 cu ft
7128/27 = 264 cu yds

Stone Ballast: Total = 26 cu yds

1st Stair –
6"/12 x 21' x 11' = 116 cu ft

1st Landing –
6"/12 x 10' x 11' = 55 cu ft

2nd Stair –
6"/12 x 18.6' x 11' = 102 cu ft

2nd Landing –
6"/12 x 3.2' x 11' = 176 cu ft

Entrance –
6"/12 x 9' x 11' = 50 cu ft
499/27 ≈ 20 cu yds + 2 Vents 6 cu yds
= 26 cu yds

Excavation: Total = 609 cu yds

1,566 sq ft. x 10.5 ft = 16,443 cu ft. / 27 = 609 cu yds

2 Vents – (11 x 15 x 4.33')/27 = 54 cu yds

Pavement Removal:

3" thick, (10.5' x 81')/9 = 95 sq yds

Concrete Removal:

(6"/12 x 10.5' x 81')/27 = 16 cu yds

Time Estimate

Concrete:

Roof — 29 cu yds × $\frac{1 \text{ day}}{10.7 \text{ cu yds}}$ = ————→ 3 days

Mat — 37 cu yds × $\frac{1 \text{ day}}{20.3 \text{ cu yds}}$ = ————→ 1 1/3 days

Walls — 46 cu yds } 74 cu yds × $\frac{1 \text{ day}}{11.2 \text{ cu yds}}$ = 7 days
Vent — 28 cu yds   }

Total = 11 1/3 days

Sheeting:

3080 sq ft × $\frac{1 \text{ day}}{295 \text{ sq ft}}$ = 10 1/2 days

---

Excavation:

609 cu yds × $\frac{1 \text{ day}}{480 \text{ cu yds}}$ = 1 1/3 days   USE 2 days

---

Backfilling:

264 cu yds × $\frac{1 \text{ day}}{235 \text{ cu yds}}$ = 1 1/8 days   USE 2 days

---

Stone Ballast:

26 cu yds × $\frac{1 \text{ day}}{160 \text{ cu yds}}$ = USE 1 day

Pavement Removal:

$$95 \text{ sq yds} \times 1.3 \times \frac{1 \text{ day}}{690 \text{ sq yds}} = \text{use } \frac{1}{2} \text{ day}$$

Concrete Removal:

$$16 \text{ cu yds} \times \frac{1 \text{ day}}{45 \text{ cu yds}} = \text{use } \frac{1}{2} \text{ day}$$

Roadway Decking:

12"×12" Timbers — $9818 \text{ B.F.} \times \frac{1 \text{ day}}{800 \text{ B.F.}} = 12\frac{1}{4} \text{ days}$

Steel —

$$2.125 \text{ tons} \times \frac{1 \text{ day}}{7.5 \text{ tons}} = \frac{1}{4} \text{ day}$$

Removal of Concrete Bulkhead:

$$28 \text{ cu yds} \times \frac{1 \text{ day}}{34 \text{ cu yds}} = \text{use } 1 \text{ day}$$

Removal of Stone Wall:

$$300 \text{ cu ft.} \times \frac{1 \text{ day}}{900 \text{ cu ft.}} = \text{use } 1 \text{ day}$$

Tunnel Flooring:

4"×8" — $760 \text{ L.F.} \times \frac{1 \text{ day}}{160 \text{ L.F.}} = 4\frac{3}{4} \text{ days}$

plywood — $1,200 \text{ sq ft.} \times \frac{1 \text{ day}}{1,350 \text{ sq ft.}} = \text{use } 1 \text{ day}$

Excavate Existing Tunnel Fill:

$$195 \text{ cuyds} \times \frac{1 \text{ day}}{6 \text{ cuyds}} = 32\tfrac{1}{2} \text{ days}$$

Electrical + Plumbing Work:

$.10 \times 81 \text{ days} = 8 \text{ days}$ say 10 days

Total = 91 working days

$$\frac{91 \text{ days}}{20 \text{ days/month}} = 4.55 \text{ months} \times 1.2 = 5.5 \text{ mos}$$
                                         ↑
                                    contingency

Say 6 months
× 2 = 1 YEAR

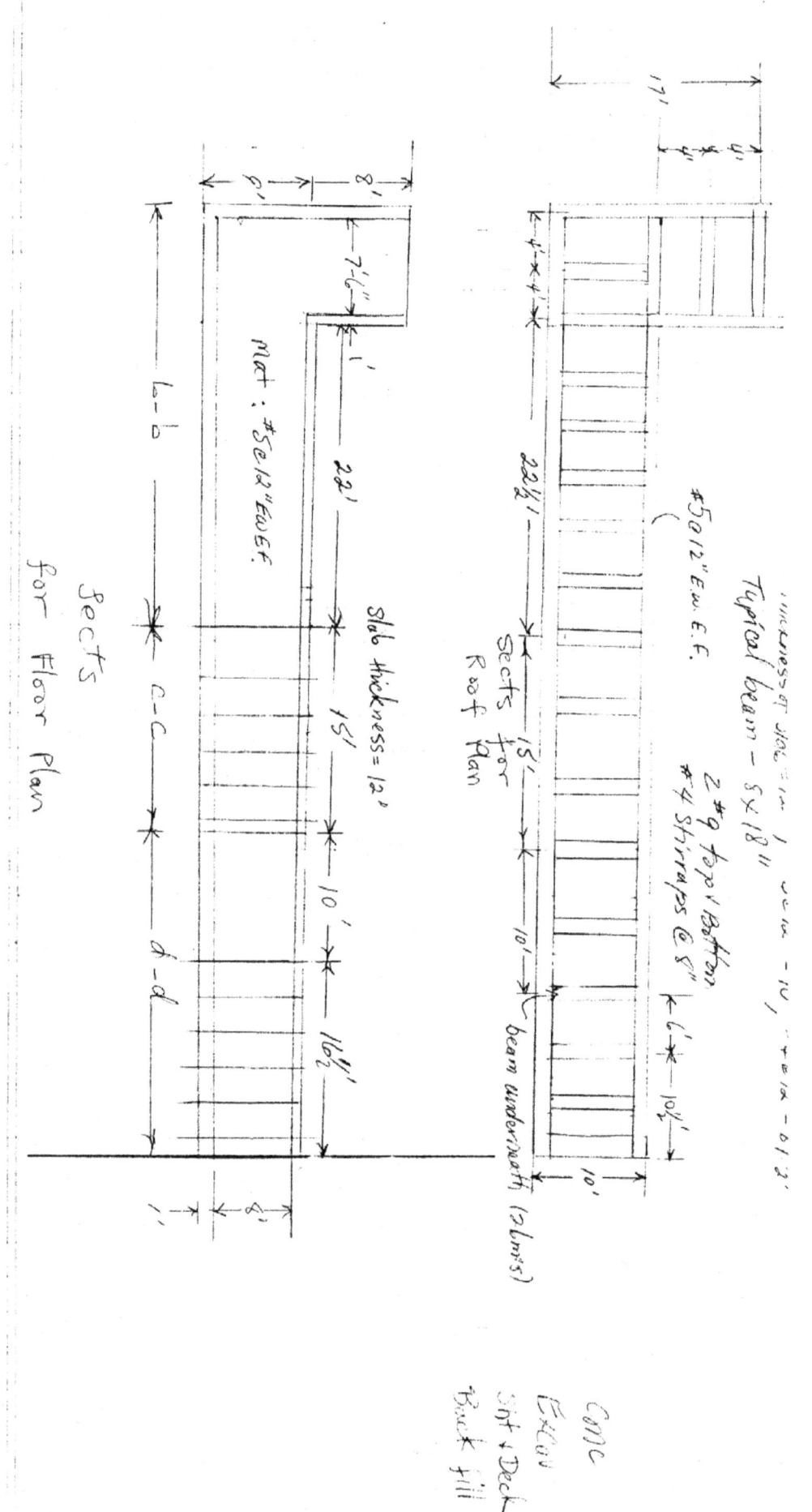

1. $w = \frac{1}{2}$ k/ft. @ 0' depth
2. $w = 1$ k/ft @ 10' depth
3. $w = 2$ k/ft @ 16' depth

$f'_c = 4000$ psi

$M_1 = \frac{wl^2}{24}$
$M_2 = \frac{wl^2}{8}$
$V = \frac{wl}{2}$

1. $w_t = \frac{1}{2}(1000/ft.) \times 8' = 4000$ lbs, $M_1 = .5^k/ft. \times 8^2/24 = 1.33$ k-ft., $M_2 = .5^k/ft. \times 8^2/8 = 4$ k-ft.
2. $w_t = 1000$ lbs/ft. $\times 10' = 10,000$ lbs, $M_1 = 1^k/ft. \times 8^2/24 = 2.67$ k-ft., $M_2 = 1^k/ft. \times 8^2/8 = 8$ k-ft.
3. $w_t = 2000$ lbs/ft. $\times 16' = 32,000$ lbs, $M_1 = 2^k/ft. \times 8^2/24 = 5.33$ k-ft., $M_2 = 2^k/ft. \times 8^2/8 = 16$ k-ft.

$V_1 = 2\sqrt{f'_c} b w d = 2 \times \sqrt{4000} \times 8/12 \times 1' = 84$ psi, $V_2 = 84$ psi.
$V_1 = 2\sqrt{f'_c}$

City Cost Index - New York - As of November 2011

The building and construction cost indexes for ENR's individual cities use the same components and weighting as those for the 20-city national indexes. The city indexes use local prices for portland cement and 2 X 4 lumber and the national average price for structural steel. The city's BCI uses local union wages, plus fringes, for carpenters, bricklayers and iron workers. The city's CCI uses the same union wages for laborers.

ENR COST INDEXES IN NEW YORK (1978-2011)

YEAR	MONTH	BCI	%CHG	CCI	%CHG
2011	Nov	7242.23	3.4	13807.28	3.0
2011	Oct	7237.23	3.5	13802.28	3.0
2011	Sep	7230.73	3.6	13795.78	3.1
2011	Aug	7229.48	3.2	13794.53	2.9
2011	Jul	7216.73	3.0	13781.78	2.8
2011	Jun	7205.23	2.9	13770.28	2.7
2011	May	7039.93	0.8	13441.53	0.4
2011	Apr	7038.93	3.2	13440.53	3.8
2011	Mar	7029.43	3	13431.03	3.7
2011	Feb	7026.43	3.1	13428.03	3.7
2011	Jan	6994.68	2.7	13396.28	3.5
2010	Dec	6998.68	2.7	13400.28	3.5
2010	Nov	7001.93	2.8	13403.53	3.6
2010	Oct	6993.43	2.7	13395.03	3.5
2010	Sep	6982.93	2.4	13384.53	3.4
2010	Aug	7008.52	2.8	13410.12	3.6
2010	Jul	7008.02	2.8	13409.62	3.6
2010	Jun	7001.52	2.5	13403.12	3.4
2010	May	6984.77	2.1	13386.37	3.2
2010	Apr	6824.18	-0.4	12951.62	-0.2
2010	Mar	6822.18	-0.4	12949.62	-0.2
2010	Feb	6816.18	-0.5	12943.62	-0.3
2010	Jan	6814.43	-0.7	12941.87	-0.4
2009	Dec	6816.18	-0.9	12943.62	-0.5
2009	Nov	6813.43	-1.7	12940.87	-0.9
2009	Oct	6813.18	-1.9	12940.62	-1
2009	Sep	6817.93	-1.9	12945.37	-1
2009	Aug	6818.68	-1.2	12946.12	-0.6
2009	Jul	6816.93	-1.1	12944.37	-0.6
2009	Jun	6829.68	2.6	12957.12	3.5
2009	May	6842.43	3.5	12969.87	3.9

Year	Month				
2009	Apr	6849.93	3.9	12977.37	4.1
2009	Mar	6851.68	4	12979.12	4.2
2009	Feb	6849.18	4.2	12976.62	4.3
2009	Jan	6863.43	4.4	12990.87	4.4
2008	Dec	6875.68	4.6	13003.12	4.5
2008	Nov	6928.68	5.5	13056.12	5
2008	Oct	6947.43	5.6	13074.87	5.1
2008	Sep	6952.68	5.7	13080.12	5.1
2008	Aug	6900.93	4.9	13028.37	4.7
2008	Jul	6894.18	4.9	13021.62	4.7
2008	Jun	6655.53	1.3	12523.87	0.7
2008	May	6613.53	0.7	12481.87	0.4
2008	Apr	6594.28	0.5	12462.62	0.3
2008	Mar	6589.28	1.1	12457.62	0.6
2008	Feb	6573.28	0.9	12441.62	0.5
2008	Jan	6573.28	0.9	12441.62	0.5
2007	Dec	6571.53	0.8	12439.87	0.4
2007	Nov	6569.03	0.5	12437.37	0.3
2007	Oct	6578.03	1	12446.37	0.6
2007	Sep	6577.03	3.6	12445.37	3.3
2007	Aug	6576.78	3.7	12445.12	3.4
2007	Jul	6574.78	3.7	12443.12	3.4
2007	Jun	6568.28	3.3	12436.62	3.1
2007	May	6567.28	4.3	12435.62	3.7
2007	Apr	6553.28	4.1	12421.62	3.6
2007	Mar	6515.56	3.6	12383.9	3.3
2007	Feb	6513.06	3.4	12381.4	3.2
2007	Jan	6513.56	3.2	12381.9	3.1
2006	Dec	6520.06	3.4	12388.4	3.2
2006	Nov	6535.31	4	12403.65	3.6
2006	Oct	6510.56	4.4	12378.9	3.7
2006	Sep	6349.46	2.5	12045.4	1.3
2006	Aug	6347.21	2.6	12043.15	1.3
2006	Jul	6340.21	2.8	12036.15	3.2
2006	Jun	6361.79	2	12057.73	2.8
2006	May	6298.56	1.1	11994.49	2.3
2006	Apr	6293.54	1.9	11989.47	2.2
2006	Mar	6290.54	2.2	11986.47	2.4
2006	Feb	6296.54	1.9	11992.47	2.2
2006	Jan	6310.51	1.9	12006.44	2.2
2005	Dec	6304.51	3.2	12000.44	2.9
2005	Nov	6281.61	2.8	11977.54	2.7
2005	Oct	6239.32	2	11935.25	2.3
2005	Sep	6194.31	0.9	11890.24	1.7
2005	Aug	6188.06	5	11883.99	7
2005	Jul	6169.29	3.3	11667.99	4.3
2005	Jun	6235.19	4.7	11733.89	5
2005	May	6230.38	4.9	11729.08	5.1
2005	Apr	6179.09	5.3	11729.08	5.8
2005	Mar	6154.37	6.2	11704.36	6.3
2005	Feb	6179.22	7.1	11729.21	6.8
2005	Jan	6193.86	8.7	11743.85	7.2
2004	Dec	6112.26	9.5	11662.25	12.3
2003	Dec	5583.09	2.7	10386.73	3.8
2002	Dec	5438.2	2	10009.06	-0.9

Year	Month	Index	%	Index	%
2001	Dec	5350.03	6.2	10101.24	7.7
2000	Dec	5018.67	-2.5	9379.14	0.3
1999	Dec	5147.21	5.3	9355.77	5.1
1998	Dec	4890.13	0.2	8899.59	1.8
1997	Dec	4880.61	2.2	8742.88	2.2
1996	Dec	4774.23	4.8	8554.47	2.1
1995	Dec	4557.44	2.2	8378.68	3.2
1994	Dec	4458.36	2.5	8117.64	4.9
1993	Dec	4349.2	4.8	7737.11	5
1992	Dec	4151.28	3.8	7367.49	3.6
1991	Dec	3997.91	3.9	7110.37	3.9
1990	Dec	3847.21	3.6	6846.49	6.1
1989	Dec	3712.2	5.4	6453.56	3.6
1988	Dec	3522.07	4.5	6231.12	4.5
1987	Dec	3369.26	4.7	5961.27	6.1
1986	Dec	3217.83	4.6	5621.15	4.3
1985	Dec	3076.19	3.1	5388.08	4.4
1984	Dec	2983.27	6.8	5160.95	5.6
1983	Dec	2792.67	7.3	4887.55	7.3
1982	Dec	2603.28	6.9	4553.93	10.4
1981	Dec	2434.62	11.3	4125.68	9.3
1980	Dec	2188.06	4.6	3774.64	5.4
1979	Dec	2091.82	11.5	3580.5	7.7
1978	Dec	1875.62	2	3325.43	7.4

Phase IIA

Design to construct an approximately 60 foot long tunnel, internal height 6'-6", internal width 6', to connect the buried steam locomotive under Atlantic Avenue, west of Hicks Street, with the tunnel interior space below Atlantic Avenue and Hicks Street.

The same concept may be used to connect the tunnel interior space with a sub-basement access point in a suitable building along Atlantic Avenue.

It is anticipated this short connecting tunnel will pass below existing underground utilities.

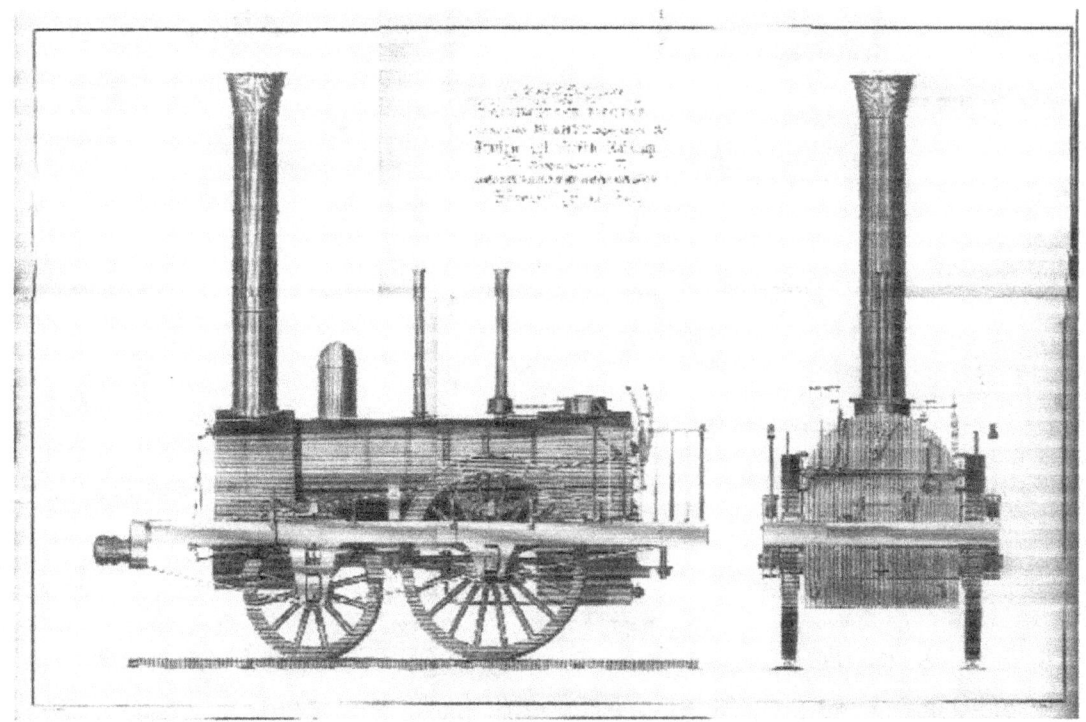

Atlantic Avenue Tunnel
Location of Iron "Anomoly"

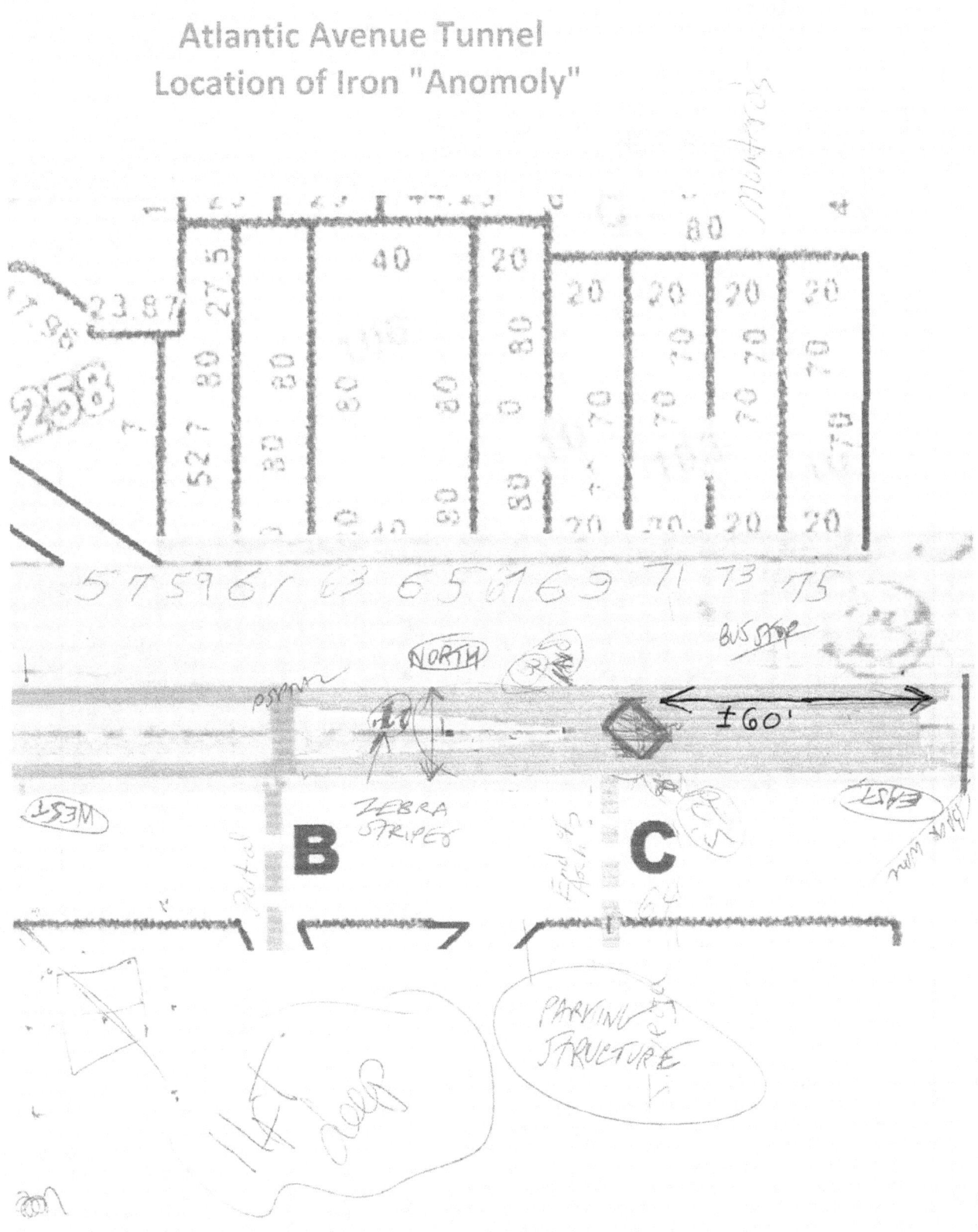

JOB: Atlantic Ave Tunnel

SOFT-GROUND TUNNELING WITH LINER PLATES 223

rib is entirely encased in concrete and may be considered as reinforcing for the permanent lining.

If the ribs are to rest on the bench, large foot plates must be welded to the bottom end to distribute the load to the footing block. Most failures of liner plates in soft ground can be attributed to foot blocks of insufficient size or to the rib slipping off the foot block. Straight ribs are to be avoided. If the walls of the tunnel are plumb, the ribs should be curved on a 200-in. radius to resist side pressure better. Plumb ribs on one section

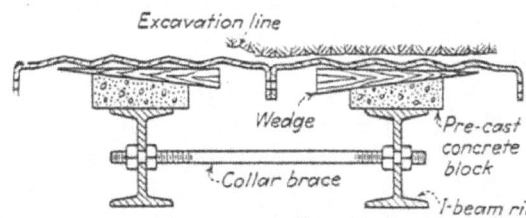

Fig. 197.—Ribs set inside of liner plates; rings of liner plates bolted to each other.

of the Chicago subway tunnels were trussed with light sets of hog rods to resist horizontal load.

SELECTING THE PLATE AND RIB

There is no way of accurately estimating the load to be carried by the primary lining. Methods have been outlined in Chap. 2 for determining the loading on the permanent lining, but it may be days or even months before this pressure develops fully. In driving through bad ground, most contractors organize to concrete every day and keep the concrete close behind the miners. In such cases, the liner plates actually support the ground from one to three days only.

The following empirical rules will serve as a guide in estimating plates and ribs required for the average tunnel in soft ground. The diameter is the excavated width of tunnel.

For tunnels 6 or 7 ft. in diameter: use $\frac{1}{8}$-in. or $\frac{3}{16}$-in. liner plates without ribs.

For tunnels 8 to 10 ft. in diameter: use $\frac{1}{4}$-in. or $\frac{5}{16}$-in. liner plates without ribs.

For tunnels larger than 10 ft. in diameter: use $\frac{1}{4}$-in. liner plates with I-beam ribs at 16-in. centers, the I-beams to be 1 in. in depth for every 3 ft. of diameter.

Beam size: W-4-13 7ft long
   2 kip/ft capacity
Insert beam thru 5" dia hole drilled in rock

Column: use 6" dia pipe, filling w/concrete. Make beam seat into pipe as follows:

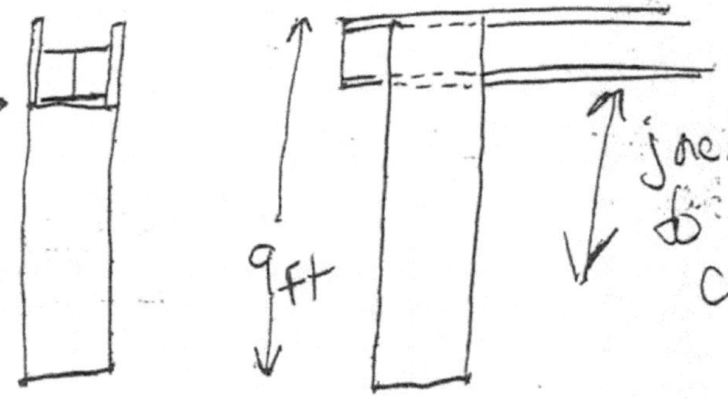

insert wedges btwn column & beam here after jacking to correct height

9ft

jack here to load-cap column

Seat pipe 3ft into floor w/grout around it.

Materials required to open wall

1 - french jack
1 - rotary impact drill 5" bit + extension shafts to drill about 4ft.
2 - bag cement
1 - W-4-13 7ft long
1 - 6" pipe 9ft long

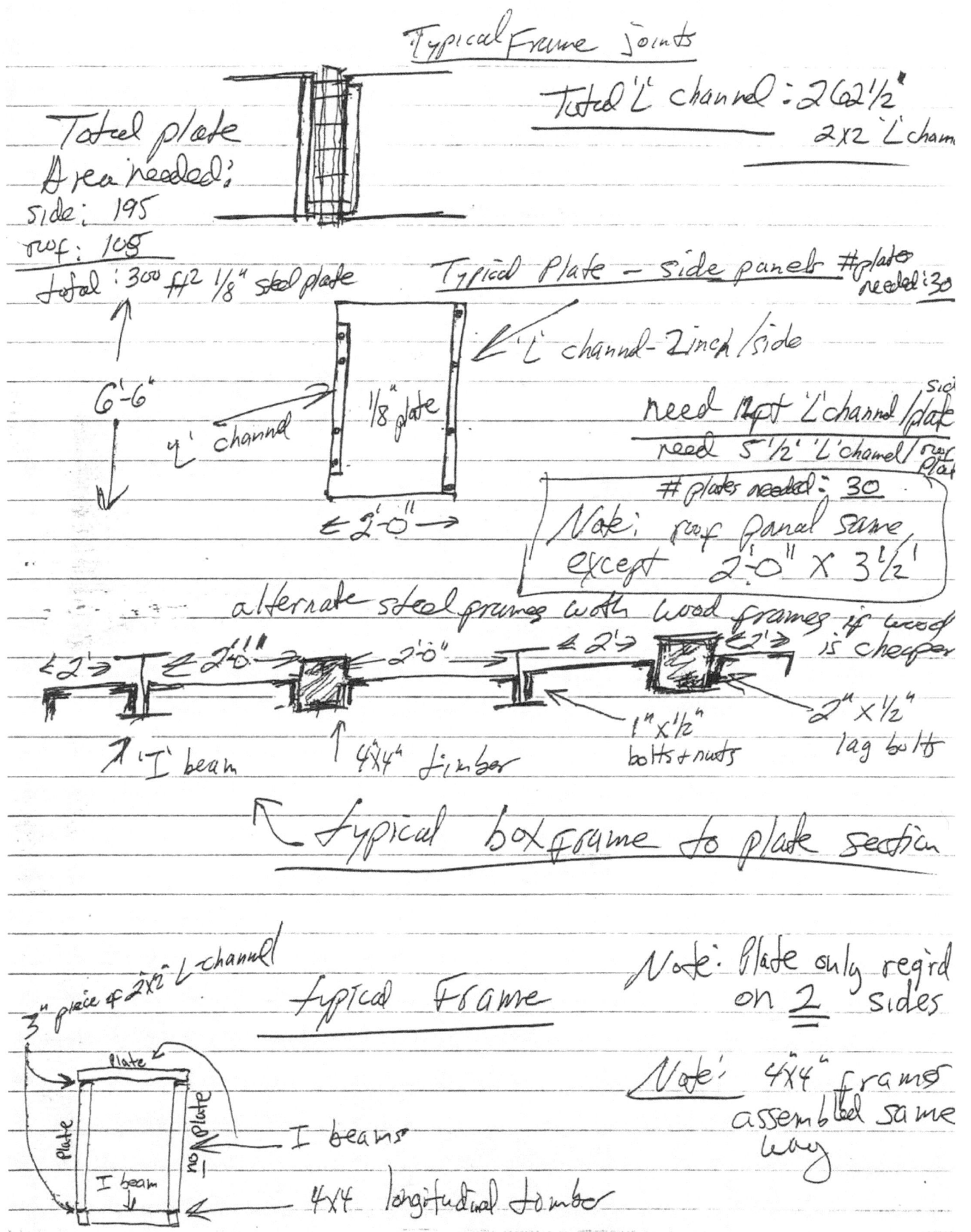

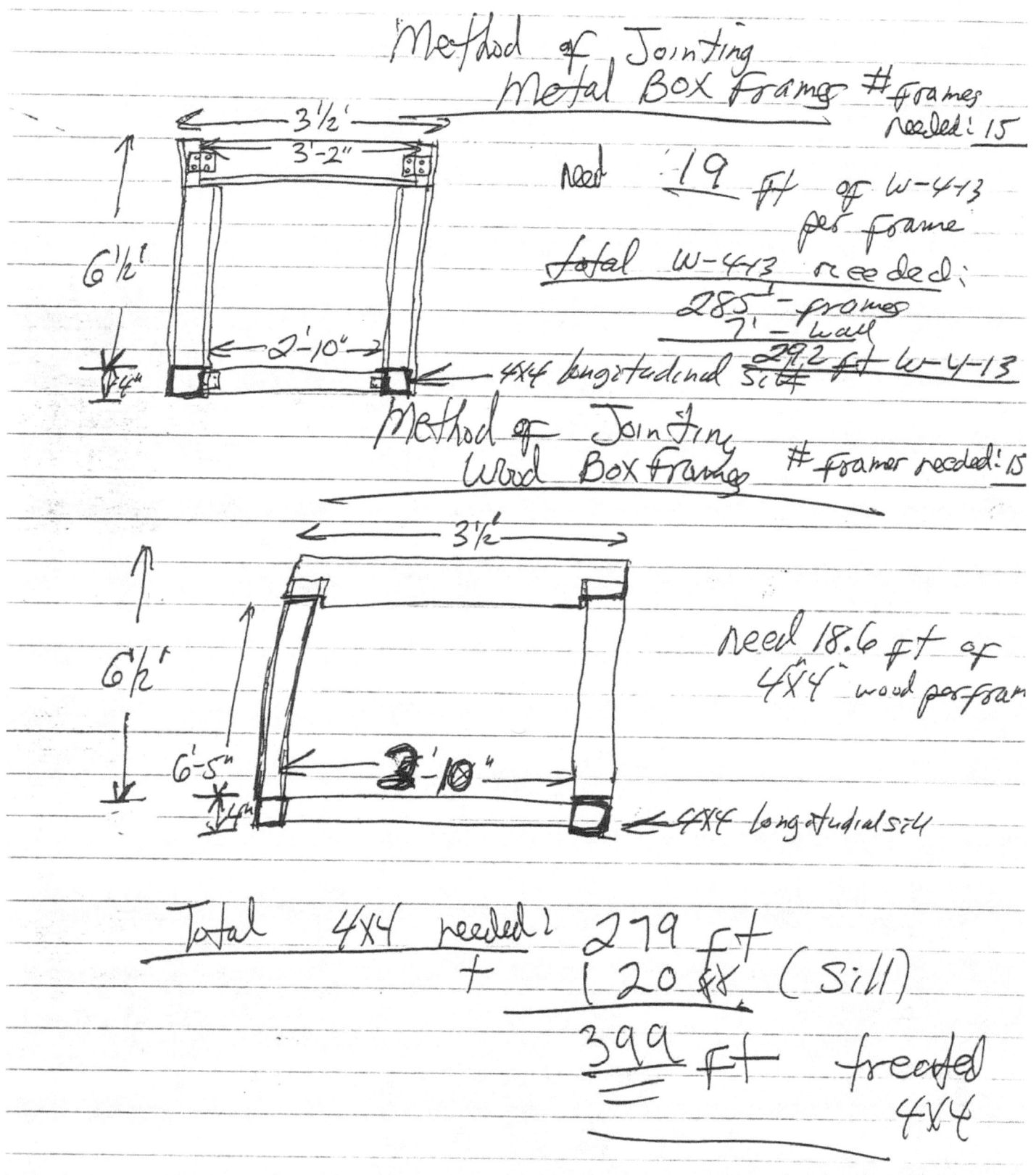

Materials Summary to go 60 feet

6" pipe — 450'

2"x2" steel 'L' channel — 262 1/2 ft

1/8" steel plate — 300 ft²

4"x4" pressure treated wood — 399 ft

W-4-13 'I' Beam — 292 ft

fasteners not included — 1/2"x1" nuts + bolts
1/2"x2" lag bolts

To Open Stone Wall

Between Stone rows #5 and #6 on right corner, Insert beam at angle thru hole drilled into sidewall knee, support other end of beam on wood or metal column. Then undermine slightly blocks directly below and remove same.

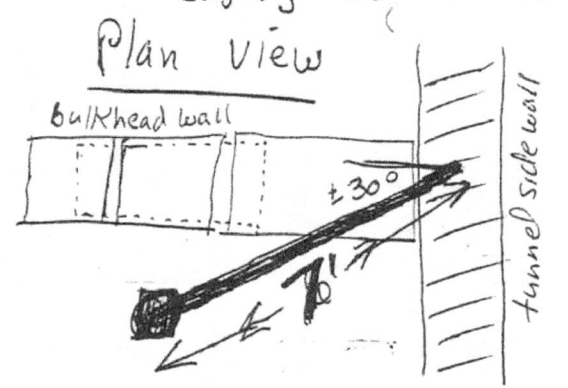

Plan view

insert beam 2ft into sidewall.

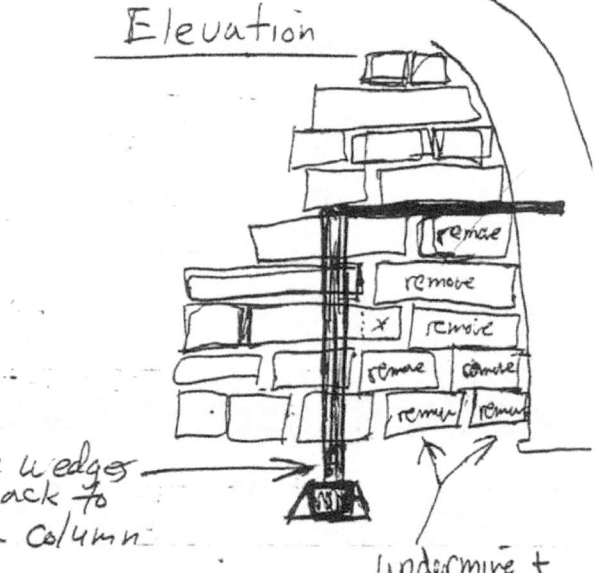

Elevation

use wedge or jack to load column

undermine + Remove

This particular point for inserting the beam was selected because the right side of the block to be supported would be held by the beam, and the left side of the block supported by cantilever action of the block under its left side. In effect, a lintel is being created.

After wall is opened, breast boards would be placed to retain fill behind wall, and drift begun.

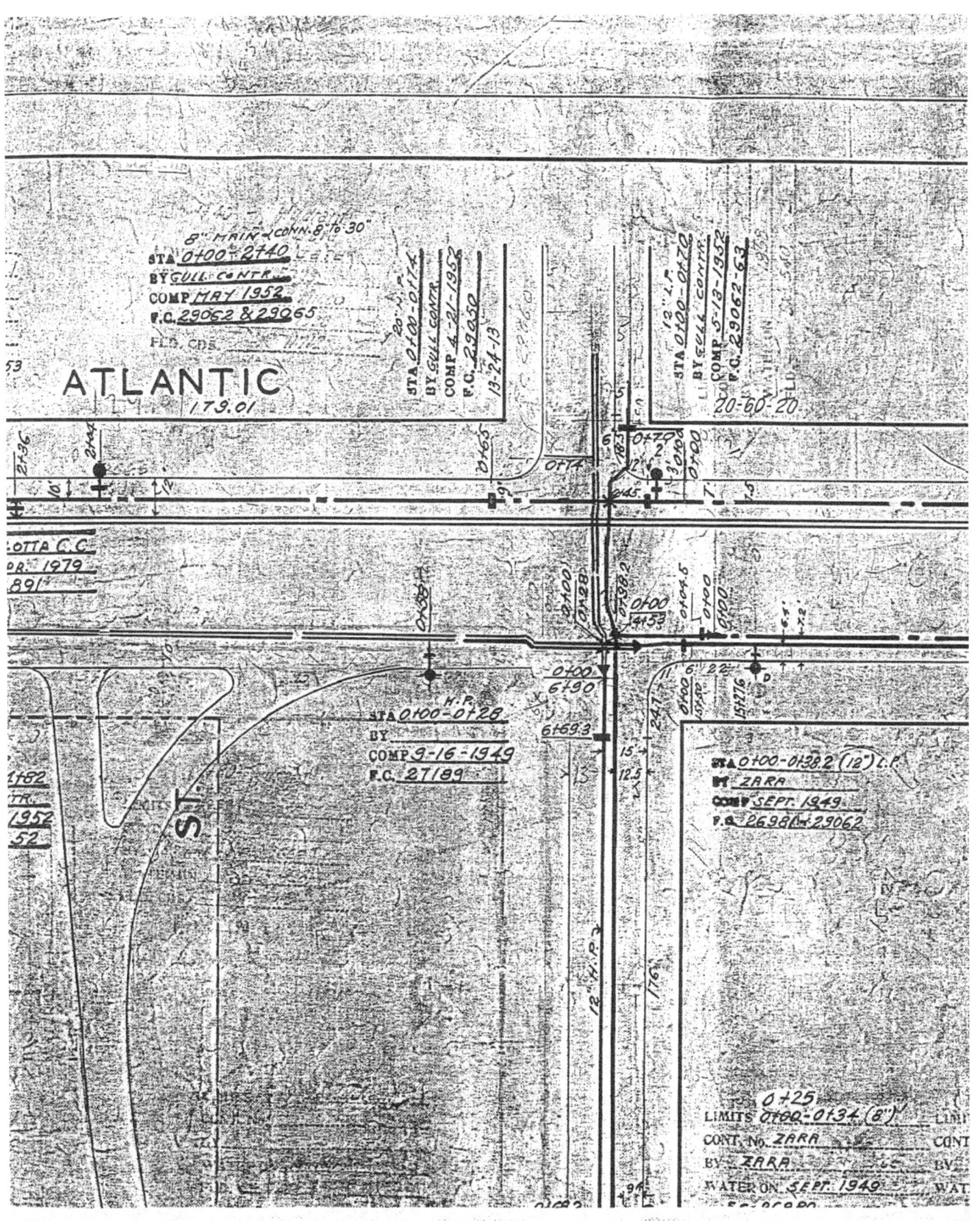

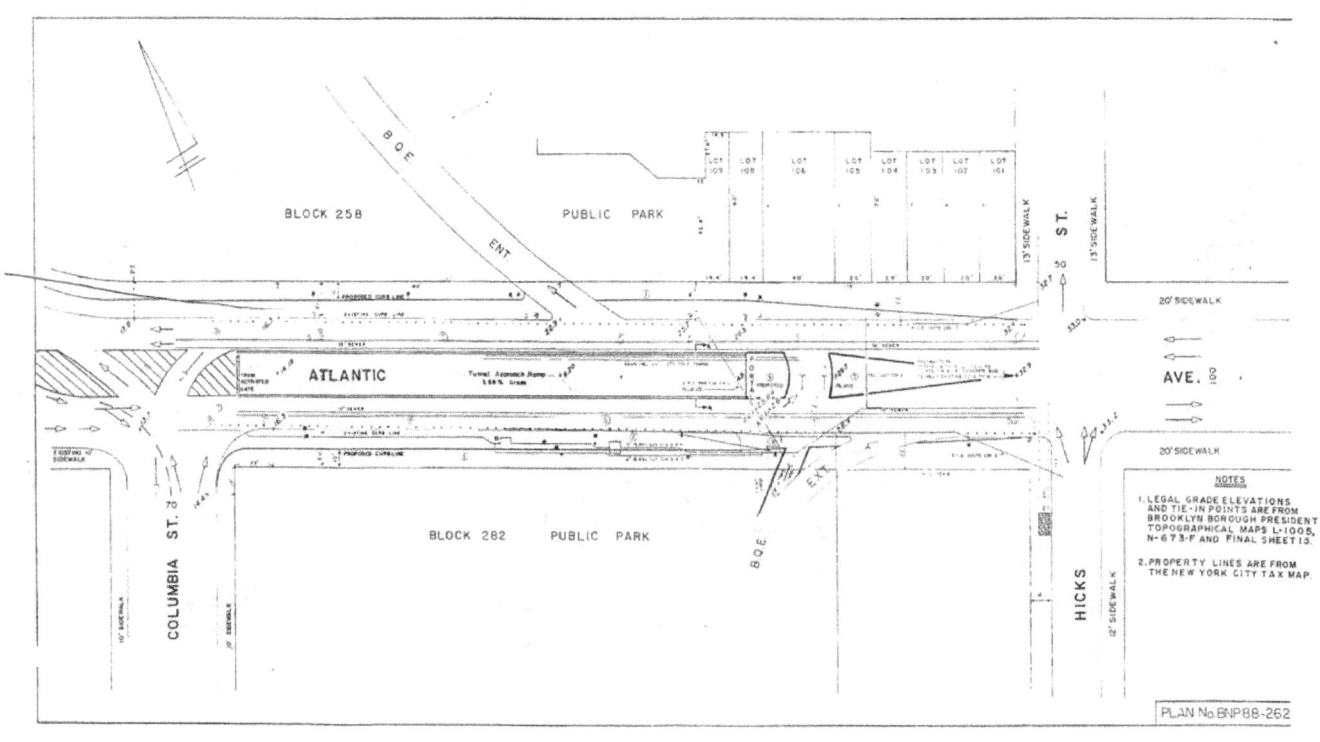

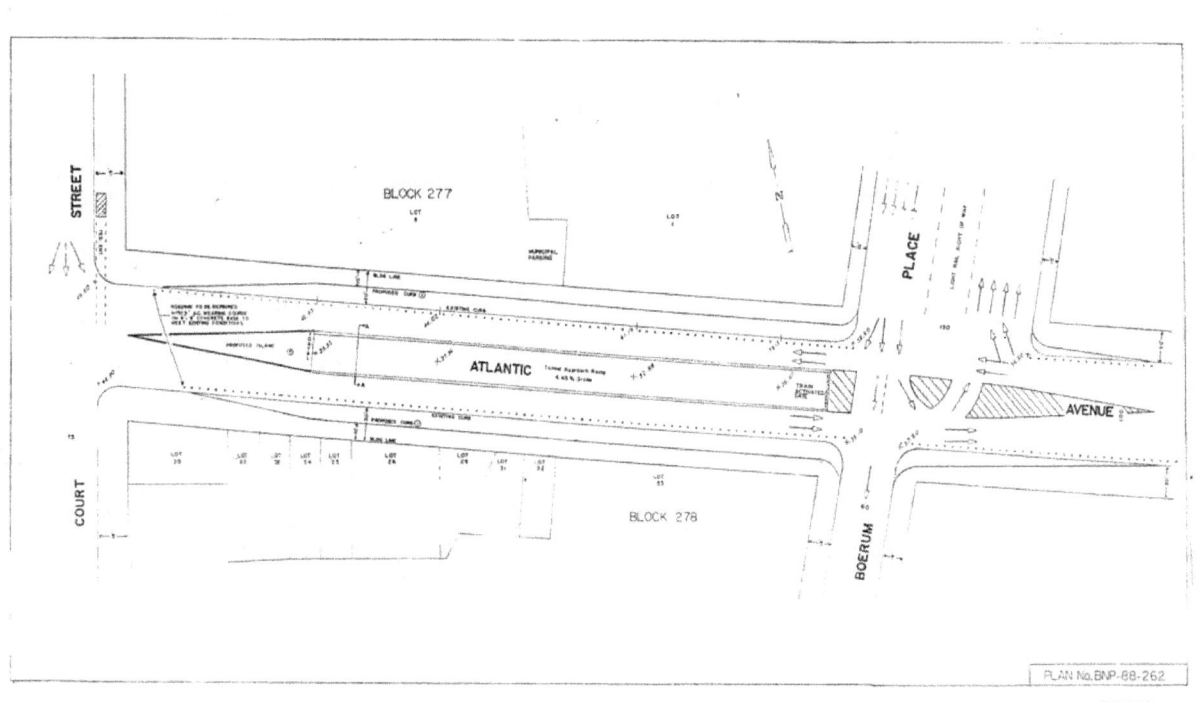

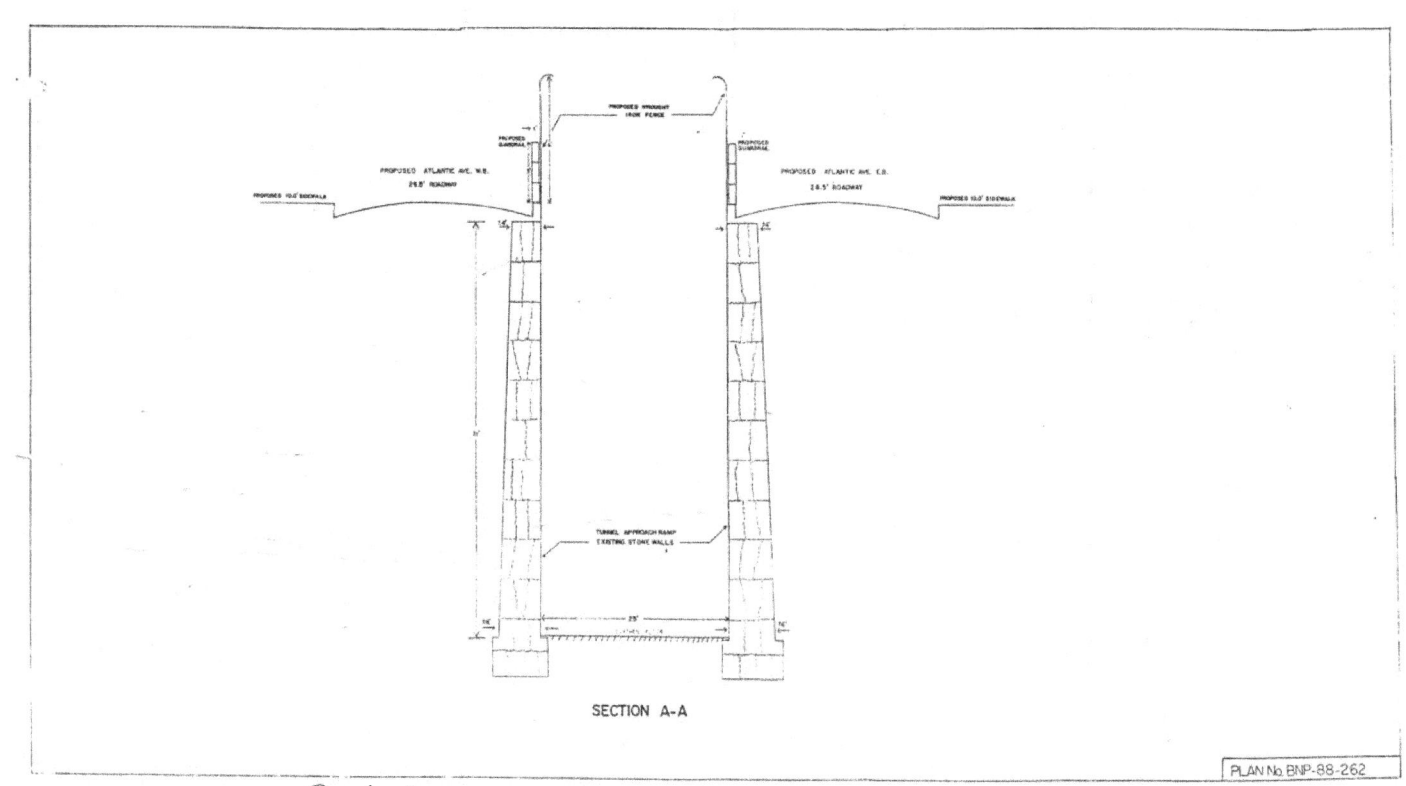

SECTION A-A

Scale: NTS

BROOKLYN HISTORIC RAILWAY ASSN.

599 E. 7th Street, Brooklyn, New York 11218
tel. 941-3160

Robert Diamond
President

Tunnel as it appeared in 1844.

June 9, 1988

Hon. Ross Sandler
Commissioner
NYC Dept. of Transportation
40 Worth Street
New York, NY 10013

Dear Commissioner Sandler:

Some time ago, our non- profit group was given a franchise to an abandoned subway tunnel under Atlantic Avenue, in Brooklyn Heights. This tunnel, was built in 1844, and sealed and abandoned in 1861. The franchise was granted by the Board of Estimate.

Recently, the City placed $2.6 million in the capital budget, for the reconstruction of the original tunnel portals. We are currently in the process of completing the final design of these entrances, and we need to confer with either yourself, or someone from your department who you designate, as these entrances are in a City street.

The tunnel will be used as a museum, and as part of a planned light rail link, between the LIRR terminal at Flatbush Avenue, and Pier 6 in Brooklyn, which would be the site of a ferry to Manhattan. The light rail would also interconnect Metrotech, Atlantic Terminal, Fulton Landing, the Brooklyn Academy of Music, and other Brooklyn sites with Manhattan.

As I mentioned, we are on a very tight schedule for the completion of the entrance design, so we would be grateful if we could here from you, or someone in your department as soon as possible. I have enclosed a very rough sketch of the proposed tunnel entrance.

Thank you in advance for your attention.

Sincerely,

Robert Diamond

cc: Jack Lusk

NEW YORK CITY
DEPARTMENT OF TRANSPORTATION
OFFICE OF THE FIRST DEPUTY COMMISSIONER

40 Worth Street   New York, N.Y. 10013

Ross Sandler
Commissioner

Samuel I. Schwartz, P.E.
Chief Engineer/First Deputy Commissioner

BN-DEM-THM-L-285

Mr. Robert Diamond
President
Brooklyn Historic Railroad Assn.
599 East 7th Street
Brooklyn, New York 11218

JUL 25 1988

RE: Atlantic Ave Tunnel
    Museum & Cultural Inst.
    Capital Project PV515
    Borough of Brooklyn

Dear Mr. Diamond:

Your letter of June 9, 1988 to Comm. Sandler is acknowledged. In order to excavate the entrance to the subject tunnel, it will be necessary for you to obtain street opening permits.

Due to the nature of your project the existing traffic patterns on Atlantic Avenue will have to be altered, consequently engineering drawings will have to be reviewed and approved by various entities in this department.

In order to begin this process of plan review and approval, please submit your site plans to Mr. Anthony Consentino, P.E., Engineer In Charge, of our Builders Pavement Section. Mr. Consentino will advise you in the required procedures in help to coordinate your project among the various divisions of Department of Transportation who must be involved. Mr. Consentino is located in room 1101 at 40 Worth Street, and can be reached at 566-3636, or 5718.

I am enclosing a copy of Highway Paving Plan for Contract BN 64-14. This widening and resurfacing of Atlantic Avenue was completed on July 5, 1966. Please note that this plan indicates that the trolley tracks over the entrance to the portal were not removed but were covered over in 1951.

We look forward to working with you on this Historic restoration project.

Very truly yours,

THOMAS H. MARKHAM, P.E.
Director-Engineering Management
Bureau of Highway Operations
Department of Transportation

Attachment:

NEW YORK CITY
DEPARTMENT OF TRANSPORTATION

OFFICE OF THE COMMISSIONER

January 13, 1989

40 Worth Street   New York, N.Y. 10013

BN-AAC-CC-HK-GFR-465

Ross Sandler
Commissioner

HARRY KAMAMIS, P.E.
Acting Assistant Commissioner

NYC Dept. of Transportation - Builder's Pavement
NYC Dept. of Transportation - Planning
Borough President - Brooklyn
NYC Dept. of Transportation - Arterial Coordinator
Consultant Engineer - Robert Diamond

RE: NYCDOT Builder's Pavement Plan
BNP 88-262
Atlantic Avenue Tunnel
Boerum Place to Furman Street
Borough of Brooklyn

Meeting to Receive Review of Proposal

Gentlemen:

Transmitted herewith ~~separately on 10 Jan. '89~~ by the undersigned ~~is~~/are 2 set(s) of the proposed plans for the above work. Your review of this submission is requested.

A meeting has been scheduled for Thursday, February 9, 1989, at 10 AM in Room 1225, 51 Chambers Street, New York, New York 10007, to receive your input and comments on this submission.

Your cooperation in reviewing the project and supplying comments is most appreciated.

Very truly yours,

GERARD F. RENINGER, P.E.
Director
Arterial Highways Coordination

Minutes of Meeting Held

Regarding Atlantic Avenue Tunnel Reopening

Interface With Other Proposed Work

Meeting Held On February 9, 1989,

10AM At 51 Chambers Street

Participants:

Gerard Reninger, Director, NYC Arterial Highways, NYC DOT
Robert Diamond, Brooklyn Historic Railway Assn.
Nayan Basu, NYS DOT Planning & Development
Jerry Blaustein, Builders Pavement, NYC DOT
Anthony Consentino, Chief, Builders Pavement, NYC DOT
Richard Pressel

The meeting was chaired by G. Reninger.
G. Reninger: The intent of this meeting is to provide an opportunity to the Dept. of Transportation and the State to comment on the effect of the reopening of the tunnel entrances as to the cost and viability of State projects near the tunnel entrances, in particular, the reconstruction of the BQE bridge over Atlantic Avenue, and the planned widening of the BQE in this area.

N. Basu: The State's concern is the BQE exit ramp at Atlantic Avenue. The State would like this particular branch of the exit to remain in its present configuration. They propse moving the western tunnel portal a short distance to the west. Mr. Basu also indicated that the narrowing of the sidewalk from 20' to 10' would slightly increase the gradient in the BQE entrance ramp on Atlantic Avenue, but felt that this would not be a problem.

R. Diamond: Mr. Diamond indicated that he would be glad to work with the State to fine tune the planning of the western tunnel portal in relation to the BQE exit ramp.

N. Basu: Indicated that he is in the process of an investigation to determine if there were any tie beams placed under Atlantic Avenue when the BQE overpass was built. He will have more information soon. So far, no tie beams have been found in the available records.

G. Reninger: Described in detail the method which will be used to reconstruct the BQE overpass at Atlantic Avenue, and the widening of the BQE. He indicated that if the west tunnel portal could not be moved further west, that a Bailey Bridge could be built over the tunnel approach ramp. DOT will need to use part of the area in the vicinity of the tunnel ramp and portal through the 1990's for their BQE project. The first project ready to go into construction must be willing to accomodate the other project. The tunnel project must be planned out so as not to preclude the BQE project, and vice versa. Mssrs. Reninger, Basu and Diamond indicated that this planning could

DOT Meeting
February 9, 1989

-2-

and would be done, and that one project would not preclude the other.

J. Blaustein: Asked if Mr. Diamond's plans would be approved by the State, if so, what would be the process.

G. Reninger and N. Basu: They stated to Mr. Blaustein that they felt that Mr. Diamond's concept for the ramp & portal near the BQE would basically work.

G. Reninger: Assured Mr. Blaustein the he and the State would work with the tunnel project. They will review the pre final plans. He will accept plans on behalf of the State for the tunnel project along Atlantic Avenue between Hicks Street and Furman Street. He wants to see that other responsible agencies have signed off also, so that the sign off from Arterial Highways and the State will be significant.

J. Blaustein: Asked what aspects of the tunnel portal plans the State will review.

N. Basu: Four areas:

1) Traffic and Safety
2) Structural
3) Design (utilities)
4) Construction

G. Reninger: Indicated that they are in the process of rehiring their BQE project design consultant, Daniel Frankfurt. Also indicated that a large diameter sewer line by cross through the west tunnel approach ramp. Daniel Frankfurt has information on the exact location of this sewer. Mr. Reninger will contact him, and provide the information to Mr. Diamond.

R. Pressel: Indicated that this sewer may already be abandoned, or may become abandoned when the Red Hook interceptor comes on line. In this case, the section of sewer through the tunnel ramp (if any) could be collapsed and bulkheaded. If the sewer does cross the ramp, and cannot be abandoned, a siphon would have to be installed, or the sewer relocated.

A. Consentino: Stated that a change in the City Map would not be required in relation to the opening of the tunnel entrances, as there will not be any change in the width of Atlantic Avenue between building lines. He stated that the changes in the sidewalk and roadway can be approved by DOT alone with a Waiver of Grade and Alignment.

A. Consentino: Stated that he is formally requesting an All Agency Conference, so that the plans for the tunnel may be reviewed by all responsible Agencies, and comments be given.

J. Blaustein: Indicated that a structural analysis will need to be performed on the tunnel approach ramp walls to determine if when excavated, they can support the load of the BQE bridge abbutments.

Respectfully Submitted,

Robert Diamond

NEW YORK CITY
DEPARTMENT OF TRANSPORTATION

40 Worth Street   New York, N.Y. 10013

Ross Sandler
Commissioner

HARRY KAMAMIS, P.E.
Acting Assistant Commissioner

SUBJECT: NYCDOT Bridges Pavement Plan
BNP-88-26Z
Atlantic Ave Tunnel Reopening
Interface with other Proposed Work.

DATE: 9 Feb 89
TIME: 10 AM
ROOM: 1225-51 Chambers St.

ATTENDANCE

NAME	TITLE	AFFILIATION & ADDRESS	TEL. NO.
GERARD F. RENINGER	Director	NYCDOT	212-566-6...
Robert S. Diamond		B'klyn Historic Railway, 599 E. 7 ST, B'klyn, NY 11218	718 941 3...
NAYAN K. BASU		NYSDOT Planning & Development, L.I.C.	718-482-45...
Jerry Blaustein		N.Y.C. DOT	566-3636

NEW YORK CITY
DEPARTMENT OF TRANSPORTATION

OFFICE OF THE FIRST DEPUTY COMMISSIONER

40 Worth Street   New York, N.Y. 10013

Ross Sandler
Commissioner

Samuel I. Schwartz, P.E.
Chief Engineer/First Deputy Commissioner

BN-DSE-ASC-C-88-262

Mr. Robert S. Diamond
599 East Seventh Street
Brooklyn, New York 11218

MAR 9 1989

RE: Atlantic Avenue Tunnel Project
Builders Pavement Plan No. BNP88-262
Borough of Brooklyn

Dear Mr. Diamond:

Our continued review of your preliminary Builders Paving Plans and the assistance which we have given you in preparing these plans to revitalize the abandoned railroad tunnel in Atlantic Avenue has now exceeded the expertise and limits of my office. We believe that this project because of its complexity must involve a comprehensive review by a number of different City, State, and possibly Federal Agencies including but not limited to a review by the private utility companies and by the community. The specialized comments and concerns of the different Agencies and groups who must become directly involved with this project are now required by my office so that their comments may be incorporated into the Paving Plans prior to consideration of final approval.

The City Planning Commission is the appropriate forum for conducting an "All Agency Conference" where the project will be discussed in depth and given further direction.

It is therefore our recommendation that you contact Mr. Andrew Karn, P.E., Assistant Chief Engineer, City Planning Commission, 22 Reade Street - Room 3N, New York, New York 10007, (212)720-3253 to assist you in filing and preparing for an "All Agency Conference."

Very truly yours,

ANTHONY S. COSENTINO, P.E.
Chief
Builders Pavement Section

www.ingramcontent.com/pod-product-compliance
Lightning Source LLC
Chambersburg PA
CBHW080916170526
45158CB00008B/2136